T0394987

Bioinformatics for biomedical science and clinical applications

Published by Woodhead Publishing Limited, 2013

Woodhead Publishing Series in Biomedicine

1. Practical leadership for biopharmaceutical executives
   J. Y. Chin
2. Outsourcing biopharma R&D to India
   P. R. Chowdhury
3. Matlab® in bioscience and biotechnology
   L. Burstein
4. Allergens and respiratory pollutants
   Edited by M. A. Williams
5. Concepts and techniques in genomics and proteomics
   N. Saraswathy and P. Ramalingam
6. An introduction to pharmaceutical sciences
   J. Roy
7. Patently innovative: How pharmaceutical firms use emerging patent law to extend monopolies on blockbuster drugs
   R. A. Bouchard
8. Therapeutic protein drug products: Practical approaches to formulation in the laboratory, manufacturing and the clinic
   Edited by B. K. Meyer
9. A biotech manager's handbook: A practical guide
   Edited by M. O'Neill and M. H. Hopkins
10. Clinical research in Asia: Opportunities and challenges
    U. Sahoo
11. Therapeutic antibody engineering: Current and future advances driving the strongest growth area in the pharma industry
    W. R. Strohl and L. M. Strohl
12. Commercialising the stem cell sciences
    O. Harvey
13. Biobanks: Patents or open science?
    A. De Robbio
14. Human papillomavirus infections: From the laboratory to clinical practice
    F. Cobo
15. Annotating new genes: From *in silico* screening to experimental validation
    S. Uchida
16. Open-source software in life science research: Practical solutions in the pharmaceutical industry and beyond
    Edited by L. Harland and M. Forster

17 Nanoparticulate drug delivery: A perspective on the transition from laboratory to market
V. Patravale, P. Dandekar and R. Jain

18 Bacterial cellular metabolic systems: Metabolic regulation of a cell system with ^{13}C-metabolic flux analysis
K. Shimizu

19 Contract research and manufacturing services (CRAMS) in India: The business, legal, regulatory and tax environment
M. Antani and G. Gokhale

20 Bioinformatics for biomedical science and clinical applications
K-H. Liang

21 Deterministic versus stochastic modelling in biochemistry and systems biology
P. Lecca, I. Laurenzi and F. Jordan

22 Protein folding *in silico*: Protein folding versus protein structure prediction
I. Roterman

23 Computer-aided vaccine design
T. J. Chuan and S. Ranganathan

24 An introduction to biotechnology
W. T. Godbey

25 RNA interference: Therapeutic developments
T. Novobrantseva, P. Ge and G. Hinkle

26 Patent litigation in the pharmaceutical and biotechnology industries
G. Morgan

27 Clinical research in paediatric psychopharmacology: A practical guide
P. Auby

28 The application of SPC in the pharmaceutical and biotechnology industries
T. Cochrane

29 Ultrafiltration for bioprocessing
H. Lutz

30 Therapeutic risk management of medicines
A. K. Banerjee and S. Mayall

31 21st century quality management and good management practices: Value added compliance for the pharmaceutical and biotechnology industry
S. Williams

32 Sterility, sterilisation and sterility assurance for pharmaceuticals
T. Sandle

33 CAPA in the pharmaceutical and biotech industries: How to implement an effective nine step programme
J. Rodriguez

34 Process validation for the production of biopharmaceuticals: Principles and best practice.
A. R. Newcombe and P. Thillaivinayagalingam

35 Clinical trial management: An overview
U. Sahoo and D. Sawant

36 Impact of regulation on drug development
H. Guenter Hennings

37 Lean biomanufacturing
N. J. Smart

38 Marine enzymes for biocatalysis
Edited by A. Trincone

39 Ocular transporters and receptors in the eye: Their role in drug delivery
A. K. Mitra

40 Stem cell bioprocessing: For cellular therapy, diagnostics and drug development
T. G. Fernandes, M. M. Diogo and J. M. S. Cabral
41 Oral Delivery of Insulin
T.A Sonia and Chandra P. Sharma
42 Fed-batch fermentation: A practical guide to scalable recombinant protein production in *Escherichia coli*
G. G. Moulton and T. Vedvick
43 The funding of biopharmaceutical research and development
D. R. Williams
44 Formulation tools for pharmaceutical development
Edited by J. E. Aguilar
45 Drug-biomembrane interaction studies: The application of calorimetric techniques
Edited by R. Pignatello
46 Orphan drugs: Understanding the rare drugs market
E. Hernberg-Ståhl
47 Nanoparticle-based approaches to targeting drugs for severe diseases
J. L. Arias
48 Successful biopharmaceutical operations: Driving change
C. Driscoll
49 Electroporation-based therapies for cancer: From basics to clinical applications
Edited by R. Sundararajan
50 Transporters in drug discovery and development: Detailed concepts and best practice
Y. Lai
51 The life-cycle of pharmaceuticals in the environment
R. Braund and B. Peake
52 Computer-aided applications in pharmaceutical technology
Edited by J. Djuris
53 From plant genomics to plant biotechnology
Edited by P. Poltronieri, N. Burbulis and C. Fogher
54 Bioprocess engineering: An introductory engineering and life science approach
K. G. Clarke
55 Quality assurance problem solving and training strategies for success in the pharmaceutical and life science industries
G. Welty
56 Advancement in carrier based drug delivery
S. K. Jain and A. Jain
57 Gene therapy: Potential applications of nanotechnology
S. Nimesh
58 Controlled drug delivery: The role of self-assembling multi-task excipients
M. Mateescu
59 *In silico* protein design
C. M. Frenz
60 Bioinformatics for computer science: Foundations in modern biology
K. Revett
61 Gene expression analysis in the RNA world
J. Q. Clement

Published by Woodhead Publishing Limited, 2013

62 Computational methods for finding inferential bases in molecular genetics
Q-N. Tran

63 NMR metabolomics in cancer research
M. Čuperlović-Culf

64 Virtual worlds for medical education, training and care delivery
K. Kahol

Woodhead Publishing Series in Biomedicine: Number 20

Bioinformatics for biomedical science and clinical applications

Kung-Hao Liang

Oxford Cambridge Philadelphia New Delhi

Published by Woodhead Publishing Limited, 2013

Woodhead Publishing Limited, 80 High Street, Sawston, Cambridge, CB22 3HJ, UK
www.woodheadpublishing.com
www.woodheadpublishingonline.com

Woodhead Publishing, 1518 Walnut Street, Suite 1100, Philadelphia, PA 19102-3406, USA

Woodhead Publishing India Private Limited, G-2, Vardaan House, 7/28 Ansari Road, Daryaganj, New Delhi – 110002, India
www.woodheadpublishingindia.com

First published in 2012 by Woodhead Publishing Limited
ISBN: 978-1-907568-44-2 (print); ISBN 978-1-908818-23-2 (online)
Woodhead Publishing Series in Biomedicine ISSN 2050-0289 (print); ISSN 2050-0297 (online)

© K-H. Liang, 2013

The right of K-H. Liang to be identified as author of this Work has been asserted by them in accordance with sections 77 and 78 of the Copyright, Designs and Patents Act 1988.

British Library Cataloguing-in-Publication Data: A catalogue record for this book is available from the British Library.

Library of Congress Control Number: 2013936995

All rights reserved. No part of this publication may be reproduced, stored in or introduced into a retrieval system, or transmitted, in any form, or by any means (electronic, mechanical, photocopying, recording or otherwise) without the prior written permission of the Publishers. This publication may not be lent, resold, hired out or otherwise disposed of by way of trade in any form of binding or cover other than that in which it is published without the prior consent of the Publishers. Any person who does any unauthorised act in relation to this publication may be liable to criminal prosecution and civil claims for damages.

Permissions may be sought from the Publishers at the above address.

The use in this publication of trade names, trademarks, service marks, and similar terms, even if they are not identified as such, is not to be taken as an expression of opinion as to whether or not they are subject to proprietary rights. The Publishers are not associated with any product or vendor mentioned in this publication.

The Publishers and author(s) have attempted to trace the copyright holders of all material reproduced in this publication and apologise to any copyright holders if permission to publish in this form has not been obtained. If any copyright material has not been acknowledged, please write and let us know so we may rectify in any future reprint. Any screenshots in this publication are the copyright of the website owner(s), unless indicated otherwise.

Limit of Liability/Disclaimer of Warranty
The Publishers and author(s) make no representations or warranties with respect to the accuracy or completeness of the contents of this publication and specifically disclaim all warranties, including without limitation warranties of fitness of a particular purpose. No warranty may be created or extended by sales of promotional materials. The advice and strategies contained herein may not be suitable for every situation. This publication is sold with the understanding that the Publishers are not rendering legal, accounting or other professional services. If professional assistance is required, the services of a competent professional person should be sought. No responsibility is assumed by the Publishers or author(s) for any loss of profit or any other commercial damages, injury and/or damage to persons or property as a matter of products liability, negligence or otherwise, or from any use or operation of any methods, products, instructions or ideas contained in the material herein. The fact that an organisation or website is referred to in this publication as a citation and/or potential source of further information does not mean that the Publishers nor the author(s) endorse the information the organisation or website may provide or recommendations it may make. Further, readers should be aware that internet websites listed in this work may have changed or disappeared between when this publication was written and when it is read. Because of rapid advances in medical sciences, in particular, independent verification of diagnoses and drug dosages should be made.

Typeset by RefineCatch Limited, Bungay, Suffolk
Printed in the UK and USA

Contents

Published by Woodhead Publishing Limited, 2013

List of figures and tables

Figures

Tables

Published by Woodhead Publishing Limited, 2013

Preface

A contemporary biomedical investigation is akin to a triathlon: multiple challenges lying ahead before reaching the final destination. Contestants are competing with other but at the same time with their own limitations. A biomedical investigation consists of at least three distinct challenges to be overcome:

1. Formulate hypotheses and investigational strategies.
2. Conduct experiments using sophisticated technology platforms and adequate materials and samples.
3. Reach a novel conclusion by systematic analysis of data.

All these challenges require a high concentration of expertise to overcome them. Although the literature offers comprehensive references and success stories, young investigators are still frustrated by the enormous amount of detail scattered throughout the many sources. The last mile of data analysis is particularly difficult to overcome. A scientist needs to dive into a deluge of complex, heterogeneous and non-linear data, remove noise, boil down the content, delineate the underlying yet obscure picture, and painstakingly piece together the fragmented pieces of the picture into an integral whole. This stage critically determines the overall value of the investigation. An insightful overview of essential knowledge would be very helpful to any young investigator, for them to accomplish their projects on time in this very competitive race.

Bioinformatics is the employment of various information technologies, together with mathematics and statistics, for accomplishing biomedical investigations. This book aims to guide young investigators and clinical practitioners during their biomedical research, releasing them from the heavy burden of trying to remember all the details from the extensive literature, by concentrating on essential knowledge, skills, success stories and available resources of bioinformatics. The content of this book is designed to be 20% theory (mostly in Chapter 1) and 80% on practical problem solving skills. Theoretical discussions are aimed at a deeper comprehension of computational methods. If they wish, readers can skip these parts and directly seek practical solutions. After reading this book, they will be able to use resources efficiently, select adequate tools for their study, and also collaborate with colleagues to design new tools when current tools fall short of their needs.

Integrated or specialized biomedical information databases, as well as associated software and tools, have been contributed by the generous bioinformatics community worldwide. These invaluable resources are largely accessible on the web. This book will enable readers to take advantage of these resources efficiently and effectively. A succinct description of useful resources is presented in each chapter. Some of the software is designed to handle tasks of heavy computational burdens. Others are lightweight versions, which offer convenient and handy solutions for viewing experimental data on a personal computer.

However, because of their constantly evolving nature, this book is not intended to serve as a user manual with step-by-step usage instructions of these software and databases, but will provide the key concepts, names, keywords and references of these resources rather than specific web links. With such information to hand, readers will be able to locate the most up-to-date version of resources using web search engines such as Google. For example, readers can locate the HapMap Genome Browser website (introduced in Chapter 2) simply by searching the phrase "HapMap Genome Browser" using Google.

How to use this book

There are a variety of ways to read this book. Readers can go through the text of each chapter, from head to toe, to obtain a comprehensive overview. Or alternatively, they can jump directly to the take home messages section, and then decide which subsections contain the solution to their current problem. One other way is to jump directly to the case studies to see how the tools are used in real examples. Finally, if the reader already has specific needs of bioinformatic resources, they can go to the boxes directly to check those available.

Of course, if what is covered in the book is still not enough, the reader can always check the citations within the text for further reading, or use key words to search the internet for the latest developments.

Published by Woodhead Publishing Limited, 2013

About the author

Kung-Hao Liang is a senior bioinformatician with more than 12 years' experience in biomedical industry and academia. He has profound experience in leading multidisciplinary bioinformatics teams. He received his Ph.D. degree in engineering from the University of Warwick in the UK in 1998. He also received postdoctoral training in Academia Sinica, Taiwan, and the Institute of Neuroscience, University of Nottingham in the UK. He was also an adjunct assistant professor in the Life Science Department of the National Taiwan Normal University. He has had many academic papers published in internationally renowned journals, in both the fields of biomedical science and computer science. He is currently a clinical scientist in the Department of Medical Research and Development, Chang Gung Memorial Hospital, Taiwan.

1

Introduction

DOI: 10.1533/9781908818232.1

Abstract: Recent progress in omics technologies, including genomics, transcriptomics and proteomics, offers an unprecedented opportunity to measure molecular activity which underlies biomedical effects. Challenges remain to transform these measurements into solid knowledge about the human body in both health and disease. Bioinformatics is the science and practice of utilizing computer technology to handle the data used to address complex biomedical problems. The reconstruction of complex biomedical systems can be seen as an inverse problem, the solution of which usually relies on assumptions derived from prior knowledge. Thus, a two-way approach of biomedical investigations from top-down, and also from bottom-up, in an adaptive fashion is briefly introduced here and further elaborated on in subsequent chapters. Practical tools, which are useful for intermediate steps during top-down and bottom-up investigations, will also be introduced.

Key words: adaptive, inverse problem, ill posed problem, omics, systems level analysis.

1.1 Complex systems: From uncertainty to predictability

Driven by curiosity, a person starts to explore their environment by using their eyes, ears and touch, immediately after birth. At this stage, these sensors are still not fully developed. Seemingly chaotic and heavy data are captured by the senses, which need to be interpreted and understood.

As a person develops and gains experience, they also learn to formulate postulation and empirical rules, and adapt and refine such rules to fit newly encountered empirical data. This is a natural adaptive and iterative process. Hypotheses which have been validated, and widely accepted by the community, then become knowledge. This new knowledge in turn will provide new angles on observations, and trigger new hypotheses for future investigation. Scientific knowledge was thus accumulated iteratively over time, exerting its tremendous impact on the course of human history.

The subject that triggers the most curiosity is about ourselves. How can we maintain and restore our health, both physically and mentally? After centuries of quests, many facets of the human body still remain elusive. Established knowledge has been obtained from well designed, hypothesis testing experiments, the cornerstones of biomedical investigations. For example, we may postulate that a certain serum protein is associated with the onset of a particular disease. This association can then be detected by experiments carefully designed to compare two specific clinical conditions (diseased vs. non-diseased) in a controlled setting (e.g. the same age group).

Obviously, a high-quality hypothesis prior to an experiment is a necessary condition for a giant leap of biomedical knowledge. Such a high-quality hypothesis often comes from novel insights derived from earlier empirical observations. But how can we achieve novel insights in the first place?

1.1.1 Theoretical issue: Ill-posed inverse problems

Challenges during the course of biomedical investigations, either biological or technical, can often be seen as inverse problems. Examples include:

- the identification of driver genes of cancer progression;
- the assembly of genomic sequences from DNA readings;
- the reconstruction of peptide sequences from tandem MS spectra;
- the detection of homologous proteins based on amino acid sequences.

An inverse problem is the opposite of a direct problem, which is to generate observable data from underlying rules. An inverse problem tries

Published by Woodhead Publishing Limited, 2013

to reconstruct the underlying rules from the data. Thus, such a problem is generally more difficult to solve than the corresponding direct problem. Usually, assumptions based on prior domain knowledge will be helpful in solving an inverse problem. As will be seen in subsequent chapters, a conceptual framework of domain knowledge is essential to guide the analysis of omics data and unveil biological mechanisms behind human health and disease.

1.1.2 Theoretical issue: Essence of biology

Biological systems have three general organizational characteristics (Hartwell et al., 1999):

- modularity;
- repeated use of the same molecular machinery;
- robustness.

Biological activities work by molecular modules, that is, some modules are activated while others are de-activated, in homeostasis or in response to outside stimuli. This modularity is the theoretical basis for module level analysis.

Biological systems were evolved by the *de novo* invention of new machinery and the repeated use of existing molecular machinery, either with complete duplication or with slight adaptation, to provide a redundant (duplicated) mechanism or to fulfil new functions. This is the theoretical basis of a variety of homological searches on genomic sequences and protein structures.

Similar biological functions may be performed by multiple molecular modules. The redundant design is a way to provide biological robustness (and stability) to the entire system.

1.1.3 Theoretical issue: Adaptive learning from biomedical data

The evolution point of view offers not only a powerful tool to conceptualize the world, but also to derive mathematical models of biological systems. Instead of scanning through an infinite search space of ill-posed inverse problems, an adaptive model may be used to generate sub-optimal solutions at greater speeds, then partly refined to better fit the data, in the same way that biological systems repetitively use existing machinery for new functions.

1.2 Harnessing omics technology

Life is complex in terms of rich and intensive information. Genetic messages not only instruct the function of cells, tissues, organs and the whole body, but also can be passed down from cell to cell and from generation to generation. Modern biomedical studies have been empowered by an arsenal of information-rich omics technology and platforms, particularly in the exploratory phases of an investigation. These technologies include, for example, next-generation sequencing technology, oligonucleotide microarray, mass spectrometry and time lapse microscopy. These technologies offer deep assays of multiple genes and proteins of clinical samples with constantly improving precision and detection limits. These technologies are also useful for the exploration of molecular mechanisms behind complex human physiology and pathology. They can lead toward the discovery of biomarkers and biosignatures for the prediction of disease onset or treatment response, a central theme of personalized medicine.

One important feature of omics technology is panoramic exploration. The panoramic exploration of biology at the DNA, RNA and protein levels is called genomics, transcriptomics and proteomics, respectively. This new feature enables new ways of study design such as genome-wide association studies (WTCCC, 2007) and genome-wide expression profiling (Eisen et al., 1998).

The high throughput nature of omics instruments inevitably creates a large amount of raw data in a short time frame. This creates great challenges in data analysis and interpretation. Levels of data complexity escalate as we ascend the ladder of central dogma of molecular biology from DNA, RNA and proteins to protein interaction networks. Mathematically, the number of interactions is factorial to the number of basic elements. For example, the pairwise interaction of 21,000 genes is C(21000,2), which is at the level of 10^8, not to mention the other interactions of more than two genes. Such large amounts of information require an innovative management of data storage, organization and interpretation. However, an adaptive strategy may be required to explore the large search space to reveal biological secrets. Informatics methodologies are thus in great demand to unleash the full power

Published by Woodhead Publishing Limited, 2013

of these instruments for basic biomedical science and clinical applications alike.

1.2.1 Theoretical issues: Non-lineality and emergence

Evidence shows that humans are more familiar with the linear way of thinking of cause and effect. Unfortunately, linear rules are difficult to characterize non-linear effects. Non-linear thinking is required to comprehend biological systems which are mostly non-linear. The current design of the human body has been polished by past evolutionary events to adapt to the environment. This adaptation is subject to non-linear interaction with multiple species and environmental factors. All biomedical phenomena seem uncertain at first sight, and empirical data are critical to sustain all potential theories.

The non-linearity of data particularly requires that we let the data speak for itself and let the rules emerge. Instead of imposing strong subjective ideas, we should sit back and listen carefully to what the data has to say. We can rely on the help of the computer by the use of adaptive models, which evolve and "fit" to the data.

1.3 Bioinformatics: From theory to practice

Bioinformatics is an evolving and exciting multidisciplinary area featuring integration, analysis and presentation of a complex dimension of omics data. The goal is to decipher complex biological systems. Computational technologies are crucial for integrating the heterogeneous bits and pieces into an integral whole. Otherwise, even very important data may seem miscellaneous and insignificant. It is also crucial to transform the data into a more accessible and comprehensible format. It is a sophisticated process with multiple levels of analysis (the platform level, the contrast level and the systems level) to reveal precious insights from the vast amounts of data generated by advanced technology.

Bioinformatics tools must be constantly updated alongside the advancement of biomedical science, as old problems are solved and new challenges continue to appear. The bioinformatics tools for cutting edge research have most likely not yet been developed. A good appreciation of the essence of bioinformatics is required so as to develop adequate tools. That is why in this book we maintain 20% of the content to theory and 80% to the practical issues. Fortunately, tools do not need to be constructed from scratch; many can be established by the augmentation of previous ones.

1.3.1 Multiple level data analysis

This book addresses the essence and technical details of bioinformatics in five consecutive chapters. Chapters 2, 3 and 4 cover genomics, transcriptomics and proteomics, respectively, the three common "omics". These chapters will introduce types of study design, major instruments available, and practical skills to condense down a deluge of data to address specific biomedical questions. Specific bioinformatics tasks are categorized into four distinct levels: the platform level, the contrast level, the module level and the systems level.

Platform level analysis is usually vendor specific, due to the variety of data formats and numerical distributions generated by various platforms. Adequate software and corresponding technical manuals have often been provided by the platform vendors for this level of analysis. In addition, a great deal of open-source software is also available with extensive testing and solid performance benchmarks by pioneering genomic centers. We will briefly discuss and provide literature references for this valuable software.

Contrast level analysis aims to detect resemblances or differences in genomic or expression patterns across different physical traits or clinical conditions. The major goal is to identify genes or biological entities, amongst all the genes in the organism, which play key roles in the biological effects under study.

Module level analysis handles groups of genes rather than individual genes, in light of a perception that biological systems operate in modules. Modules may manifest as co-expressed genes or interacting proteins. Different molecular modules are activated whilst others are deactivated in response to internal changes or external stimuli. The activated proteins may interact with each other to fulfill their functions.

Finally, systems level analysis features a joint extensive analysis in two or more types of assays (e.g. DNA + RNA), which can render deeper cause-and-effect insights into biomedical systems.

Chapters 5 and 6 cover systems biomedical science (SBMS) and its clinical applications. SBMS is a further extension and abstraction of the

Published by Woodhead Publishing Limited, 2013

systems level analysis covered in the previous chapters, aiming to adequately integrate heterogeneous data to draw novel conclusions from a holistic point of view. The major emphasis is on an adaptive top-down and bottom-up model for the exploration of uncertain space and for capturing the implicit information from the data. Chapter 6 addresses the challenges to transform knowledge from exploratory investigations into clinical use. Case studies are provided in all main chapters, demonstrating how the aforementioned techniques were used flexibly in an integral fashion to unveil novel insights.

Although we categorize the content into genomics, transcriptomics and proteomics, many contrasting levels and module level analysis can be shared across platforms, with slight adaptations, as long as the underlying assumptions (i.e. data normality) are met.

1.4 Take home messages

- General or specific bioinformatics tasks are often ill-posed inverse problems, requiring prior knowledge or conceptual frameworks to solve them.
- Adaptive models are important for finding solutions in an infinite search space.

1.5 References

Eisen, M.B., Spellman, P.T., Brown, P.O. and Botstein, D. (1998) Cluster analysis and display of genome-wide expression patterns. *Proc. Natl. Acad. Sci. USA*, 95: 14863–8.

Hartwell, L.H., Hopfield, J.J., Leibler, S. and Murray, A.W. (1999) From molecular to modular cell biology. *Nature*, 402: C47–C52.

WTCCC (2007) Genome-wide association study of 14,000 cases of seven common diseases and 3000 shared controls. *Nature*, 447: 661–78.

2

Genomics

DOI: 10.1533/9781908818232.9

Abstract: The human genome is the blueprint of the human body. It encodes valuable yet implicit clues about our physical and mental health. Human health and diseases are affected by inherited genes and environmental factors. A typical example is cancer, thought to be a genetic disease, where a collection of somatic mutations drives the progression of the disease. Tracking the disease roots down to the DNA alphabet promises a clearer understanding of disease mechanisms. DNA alphabets may also serve as stable biomarkers, suggesting personalized estimates of disease risks or drug response rates. In this chapter, we explore techniques critical for the success of genomic data analysis. Special emphasis is placed on the adaptive exploration of associations between clinical phenotypes and genomic variants. Furthermore, we address the skillful analysis of complex somatic structural variations in the cancer genome. An effective use of these skills will lead toward a deeper understanding and even new treatment strategies of various diseases, as is evidenced by the case studies.

Key words: genome variants, sequence alignment and assembly, genome-wide association studies, multiscale tagging, Manhattan plot, haplotype, adaptive models.

2.1 Introduction

The human genome underlies the fundamental unity of all members of the human family, as well as the recognition of their inherent dignity and diversity. In a symbolic sense, it is the heritage of humanity.

(*Universal declaration of the Human Genome and Human Rights*)

The human genome, the entire inherited blueprint of the human body, physically resides in every nucleated human cell. A rich content of genomic information is carried by the 23 pairs of autosomal and sex chromosomes, as well as the mitochondrial genome. Chemically, the human genome is a set of deoxyribonucleic acid (DNA) molecules, comprised of long strings of four basic types of nucleotides: adenine (A), thymine (T), guanine (G) and cytosine (C). The human genome comprises 3.2 billion nucleotide base pairs, where a base pair is either a G-C pair or an A-T pair, connected by a hydrogen bond. Genes and regulatory elements are encoded by the four nucleotides. A gene refers to a DNA sequence that can serve as a template for the assembly of another class of macromolecules, ribonucleic acid (RNA). The protein-coding messenger RNAs are produced according to the DNA templates in a process called transcription, which is also controlled by the upstream DNA sequence of the gene called the regulatory element.

Genomic DNA is largely identical among healthy nucleated cells, except for certain specialized immune cells and some harmless somatic mutations scattered throughout the genome. Therefore, it is accessible in many types of tissues of the human body. DNA has a double helical structure with two strands of DNA tightly bound by hydrogen bonds. As a result, DNA is a more stable molecule and more static than downstream mature RNAs (mRNA) and proteins, which are dependent on time, tissue expression and degradation patterns. Thus, it is a reliable material for the study of the molecular basis of clinical traits.

A haploid human genome (i.e. the DNA in one strand of the double helix) comprises 3.2 billion bases. Despite its sheer length, only 28% of it harbors the 21,000 protein-coding genes (IHGSC, 2004). The rest of the genome encodes regulatory functions, contains repetitive sequences (>45%), or plays unknown roles waiting to be discovered. For example, a promoter region preceding a gene can serve as the site where the polymerase binds, regulating the transcription of this gene. The precursor RNA transcripts, directly transcribed from the DNA template, are subject to a splicing process before the mRNA is produced. Most of the precursor regions, called introns, are spliced out. As a result, only a small fraction (called exons, covering 1.1–1.4% of the genome) contribute to the final mRNA, which is further translated into proteins. Metaphorically, these protein-coding genes are

Published by Woodhead Publishing Limited, 2013

analogous to archipelagos scattered in the wide open sea of the entire genome. The exonic regions are like the islands in an archipelago, which can give rise to mRNAs. The intron regions are analogous to the inner sea. The exons are then further categorized as 5'untranslated regions, protein coding regions, and the 3'untranslated regions, depending on whether or not they directly encode the protein sequence.

2.1.1 Global visualization by genome browsers

A sea is so large that even an experienced sailor needs the navigating tools of maps and compass to direct his exploration. Similarly, the deluge and richness of genomic information prohibit investigators from comprehension and analysis without the aids of computational tools. Computers are not only used to store, organize, search, analyze and present the genomic information, but also to execute adaptive computational models to illustrate the characteristics of the data. Visual presentations are crucial for conveying the rich context of genomic information in either the printed forms or the interactive, searchable forms called genome browsers. Here we introduce three commonly used diagrams for genomics. First, a karyogram (also known as an idiogram) is a global, bird's eye view of the presentation of the 23 pairs of chromosomes and their annotations. For example, it can be used to present the tumor suppressor genes by showing their locations on the chromosome. One of the popular tools for producing a virtual karyogram is provided by the University of California, Santa Cruz (UCSC) Genome Bioinformatics site (Box 2.1).

Box 2.1 Bioinformatics resources and tools for genomics

A. Integrated Genomic Database

- NCBI GenBank

GenBank, hosted in the NCBI (National Center of Biotechnology Information), is one of the most valuable resources of genomic information. It provides DNA, RNA and protein sequences through a spectrum of specialized databases. Some are designed with extensive curation aiming to provide only high-quality data (e.g.

RefSeq). Others are focused on wide incorporation of exploratory data. The dbSNP database contains information about known SNPs. The NCBI site is one major place for newly discovered sequences to be submitted and deposited. The accompanying search engine is called Entrez, where users can locate information using sequence accession numbers (IDs) and key words. The default sequence alignment algorithm is BLAST, an algorithm combining heuristic seed search and dynamic program for alignment extension. NCBI also offers several valuable tools for experiment design, such as:

- Primer-BLAST

This is a web service or primer design based on primer 3, one of the most commonly used primer design tools. The Primer-BLAST has added-on, multiple and practically useful functions such as intron inclusion (an ideal feature to design primers for alternative splicing), exon-exon boundary check or exon-intron check. It also checks all the possible amplicons (PCR products) for a pair of primers from the database, including amplicons generated from the imperfect base pairing of the primers to the templates. This is a useful feature for the design of the primer, to avoid primers that can generate unwanted products.

- e-PCR

This website is used to check the expected amplicon of a pair of primers. This is based on the blast sequence alignment against the built-in genome and transcriptom database (i.e. RefSeq). This tool can be used to check whether the designed primer pair results in a multiple hit, and check whether the primer pairs will produce contamination, which is not to be pursued.

- UCSC Genome Bioinformatics

The UCSC Genome Bioinformatics site, hosted by the University of California Santa Cruz, contains complete and organized genomics data as well as many levels of annotations. It is best known for an interactive genome browser. It also provides a variety of information

Published by Woodhead Publishing Limited, 2013

query methods. Furthermore, the sequence alignment tool BLAT is very efficient for database queries of homologous sequences.

- HapMap

HapMap is a series of population genomic studies of several ethnic groups worldwide. The website contains valuable information about SNPs and population specific allele frequencies. An interactive genome browser is provided on the website. Furthermore, Haploview (provided by the Broad Institute of MIT and Harvard) is a standalone software for viewing the haplotype data from HapMap.

- SeattleSNPs

SeattleSNPs, also known as the Genome Variation Server, offers a convenient online portal site for presenting, browsing and searching of SNP-related information such as LD. HapMap data can also be accessed here.

- Genecards

Genecards provides extensive annotation and information about human genes, gene symbols, their functional descriptions, corresponding RNA IDs and isoforms, as well as protein information.

- OMIM

OMIM stands for Online Mendelian Inheritance in Man. It offers high-quality annotations on individual genes and their mutations associated to genetic diseases.

B. Tools for genome analysis

- CIRCOS

CIRCOS is a Perl-based platform useful for generating circular chromosomal maps. This software has been used extensively to generate high resolution, publication-quality presentation of genomic studies, such as somatic mutations or chromosomal rearrangements.

- GenomeBrowse

GenomeBrowse is a standalone software offered by the Golden Helix Inc.

- ALLPATH, ALLPATH-LG and Velvat

ALLPATH and ALLPATH-LG offers *de novo* assembly tools for NGS readings. Velvat is also a *de novo* sequence assembler of NGS readings, hosted by EMBL-EBI.

- Bowtie

Bowtie software offers the mapping of NGS readings of the long sequences such as the human genome. This software is downloadable from sourceforge.net. There are two major versions: Bowtie 1 (Langmead et al., 2009) and Bowtie 2 (Langmead and Salzberg, 2012). The latter offers local alignment and gapped alignment. Bowtie 2 is also faster and more sensitive than Bowtie 1, when the average sequence reading is larger than 50 bases.

- Chromas Lite

Chromas Lite is a lightweight software for showing the analog chromatogram signals in the trace files and also produces base calls.

- NovoSNPs

NovoSNPs is a lightweight, standalone software for detecting SNPs from several chromatogram traces.

- THESIAS

THESIAS offer JAVA-based standalone software for phasing genomic variants and then followed by haplotype vs. phenotype associations.

- UGene

UGene is a free, standalone package for sequence analysis. It can be used for showing sequence trace files, six frame translation of nucleotide sequence, multiple sequence alignment and primer design.

C. Visual presentation and interpretation

- WGAviewer

WGAviewer is a standalone software, provided by Duke University, to present a series of GWAS P-values in Manhattan plots and also

Published by Woodhead Publishing Limited, 2013

annotate the most significant SNP hits by connecting to online databases such as Ensemble and HapMap. The presentation is also facilitated by ideogram, Q-Q plot, the r^2 values of a SNP with adjacent SNPs, and gene locations. The software is written in the JAVA programming language.

- LocusZoom

LocusZoom is a useful website service, provided by the University of Michigan, for presenting regional SNP association results with annotations (Pruim, 2010). This site incorporates data from several GWAS for demonstration and re-analysis. Users can also submit their own data for analysis and presentation. The default input format is PLINK. A stand alone version of the software, built upon Phython and R, is also available for linux platforms.

- SNAP

SNAP stands for SNP Annotation and Proxy search offered by the Broad Institute. It offers an LD plot, which can also show recombination sites. Data of the HapMap and the 1000 genome project are incorporated in the site. This can be used for the imputation of associations on unassayed variants, by the use of LD from the public domain data.

- Octave

Octave is a free scientific computation package. It employs the scripting language of Matlab, another widely used commercial scientific computation package. Octave and Matlab are excellent for matrix and vector computations. They are also useful for visual presentation of 2-D and 3-D scientific data.

Second, a genome browser is an integrative, interactive visualization tool that can present genomic information, chromosome by chromosome, with multiple resolutions of annotations. Several genome browsers are part of important bioinformatic portals, for example, the Ensemble and HapMap websites (Box 2.1). A genome browser can also be standalone software, for example, the freeware "GenomeBrowse" offered by Golden

Helix Inc. A genome browser usually has an interactive, zooming in and zooming out interface to navigate to regions of interest and adjust to the right scale/resolution. This enables the visualization of large-scale contexts and small-scale details. It also has multiple tracks, which can each be designated to a special type of annotation such as the exon and intron locations, personal variants (to be discussed later), splicing patterns and functional annotations. The presentation of multiple tracks simultaneously offers an integrative view of complex information.

A genome browser is traditionally designed as a linear, horizontal interactive graph with multiple layers of annotation showing simultaneously as tracks. With the current requirement of presenting complex gene interactions and structural variations across chromosomes, a new type of genome browser has been designed with a circular shape with the chromosomes forming the circle. Annotations are then represented in the circular tracks, and the interactions of loci are shown as arcs within the circle. Recently, such a type of genome diagram and browser was constructed by the CIRCOS software (Box 2.1).

2.2 The human genome and variome

2.2.1 Genomic variants

A substantial portion of the human genome is identical amongst people; the differences are called variants. Many such variants are scattered in the human genome. They can be classified, based on their lengths (scales), as:

- microscopic variants (>3 mega bases, 3 M);
- sub-microscopic variants, such as copy number variations (CNVs); and
- small variants (<1 kilo bases, 1 kb), including single nucleotide polymorphisms (SNPs), point mutations (single-base substitutions), short tandem and interspersed repeats (Feuk et al., 2006).

The collection of all variations found in the human population is called the human variome.

The single-base SNPs and point mutations are important classes of genomic variants, partly because of their sheer numbers. Large-scale human genome sequencing projects, such as the 1000 Genome Project and the uk10k Project (a 6000 vs. 4000 case vs. control study aiming to

Published by Woodhead Publishing Limited, 2013

conduct the sequencing of the protein-coding regions), have increased the SNP count to 20 million (Pennisi, 2010). Both SNPs and point mutations have two (occasionally three) alleles: the more common one is called the major allele, while the less common one is called the minor allele. A SNP is defined as having a minor allele frequency (MAF) higher than 1% in a population. The International HapMap Project has achieved a valuable database that can be used to calculate allele frequencies across populations (The International HapMap Consortium, 2007; Altshuler et al., 2010). The HapMap Project is prioritized to collect common variants with MAF higher than 1%. The core of the database is a large SNP-by-subject matrix of genotypes. The Phase 3 collection contains genotypes of 1,600,000 SNPs of 1184 subjects in 11 populations. The database has many uses, such as the comparison of allele frequencies in different ethnic groups and the demonstration of genomic structure based on correlations between SNPs.

On average, there is 1 SNP in every 150 nucleotides (20 million SNPs scattered in a 3 billion nucleotide genome, see Table 2.1). Hence, SNPs offer dense genomic markers. Each SNP has a unique position in the genome and is assigned a unique identification (ID) by dbSNP, one of the most important resources of SNPs hosted by the National Center for Biotechnology Information (NCBI) site (Box 2.1).

In addition to SNPs and point mutations, short tandem repeats and interspersed repeats are commonly found in the human genome. Alu is a class of primate specific, short interspersed repeats, which constitutes approximately 10% of the human genome (Lander et al., 2001; Decerbo

Table 2.1 **Information of various projects and technological platforms measured by numbers of nucleotide bases**

Item	Bases
The human genome	3,000,000,000
Known total SNPs	20,000,000
Estimated number of SNPs with MAF >10%	7,000,000
HapMap Phase 3 SNPs	1,600,000
Affymetrix 6.0 SNPs	900,000
Illumina HumanHap 650Y	650,000
Illumina HumanHap 550	550,000
Affymetrix 5.0, 500k	500,000

and Carmichael, 2005). Cellular roles of Alu have been obscure and it was once thought of as junk DNA (Muotri et al., 2007; Lin et al., 2008). Alu can be found in intergenic regions as well as the coding and non-coding regions of many genes in either the sense or antisense strand.

Recent studies on human genome analysis have revealed a common class of germline variants, the CNVs that belong to a more general class called structural variants. The genome is thought to be diploid where one copy is from the father and one from the mother. Yet, on the sub-microscopic scale, multiple genomic regions exhibit multiple elevated (e.g. above 3 copies) or reduced (e.g. 1 or 0 copies) numbers of copies, and are called CNVs. These are common, mostly harmless variants, although some have been linked to disease. The shortest length of a CNV is 1 kb based on its definition, yet many of them are longer than 100 kb. It is estimated that an average Asian person has 20 to 30 CNV loci in their genome, with an average size of 14.1 kb (Wang et al., 2007).

Cohorts of subjects need to be recruited to reveal the basic characteristics of genomic variants, such as total number and population frequency. The HapMap Project is a good example. In addition, we also need to find the genomic variants responsible for individual variability in disease susceptibility or drug response. This can also be achieved by genomic studies with adequate sample size, rigorous statistical measures, and unambiguous definition and recruitment of subjects with distinct clinical phenotypes.

Somatic mutations of various scales play significant roles in the etiology of many illnesses, particularly cancer. Most cancers originated from somatic cells transformed via a series of genetic or epigenetic modifications during the aging process (Luebeck, 2010; Stephens et al., 2011). These cells gradually exhibit many abnormal capabilities such as anti-apoptosis, tissue invasion and metastasis (Gupta and Massague, 2006). Somatic mutations of oncogenes or tumor suppressor genes could serve as important biomarkers in various stages of cancer biology. In terms of treatment, traditional chemotherapy agents have been designed against fast dividing cells. The more recent target therapies employ antagonists to various oncogenes directly. Cancer patients show varied responses, chemotherapy or targeted therapy alike, because their cancer cells are at different stages and with various features and capabilities. This is why the optimum treatment strategies differ. Studies of somatic mutations on targeted genes, oncogenes and tumor suppressor genes may be able to identify the best treatment strategy for each individual.

Published by Woodhead Publishing Limited, 2013

2.2.2 A multiscale tagging of your genome

Linkage disequilibrium (LD) is a unique feature among genomic variants. It refers to the association of genotypes between adjacent variants. It is a direct consequence of chromosomal recombination (also known as crossover), which occurs during meiosis for the production of germ cells. Where there are no chromosomal recombinations, all variants in one chromosome should be tightly associated to each other, as they are passed down together to future generations. A chromosomal recombination disrupts the association (and introduces independence) between variants at the two sides of the recombination spot. After several generations, distant variants are no longer associated due to many recombination hotspots between them. The adjacent variants still retain a certain degree of LD, particularly those between two recombination hotspots. LD of a pair of variants can be quantified by r^2 or D' values. r^2 is a value between 0 and 1, while 0 indicates no association and 1 indicates high association. LD can be shown as LD-heat maps by the Haploview software or the SeattleSNPs site (Box 2.1).

LD is a cornerstone for the design of genotype vs. phenotype association studies. Since many variants are adjacent to the causative variant, their genotype may also be in LD with the causative genotype. In the results of many high density association studies, we have seen multiple adjacent SNPs associated to the clinical trait simultaneously. The concept of using adjacent variants to serve as surrogates can alleviate the burden of finding the exact causative variants directly. We can first detect the surrogates, then scrutinize the proximity region of the surrogate to find the real causative variant. In practice, this is realized by the selection of representative SNPs (known as tag SNPs) in the study design phase, assuming the true causative variant is in LD with the tag SNPs.

A variety of tagging software is available based on r^2 LD statistics (Box 2.1). For example, a tag SNP selection software, called the LDSelect method, has been implemented for data organization and analysis on the SeattleSNPs website (Carlson et al., 2004). LDSelect can ensure the selected tag SNPs have relatively low LD between each other, thereby minimizing the number of SNPs while maximizing their representativity. The FastTagger is a GNU GPL open source software (Liu et al., 2010). It implements two tagging algorithm, MultiTag and MMTagger. FastTagger comprise two consecutive steps:

1. merging SNPs with 100% r^2;
2. tagging SNPs subject to a given r^2.

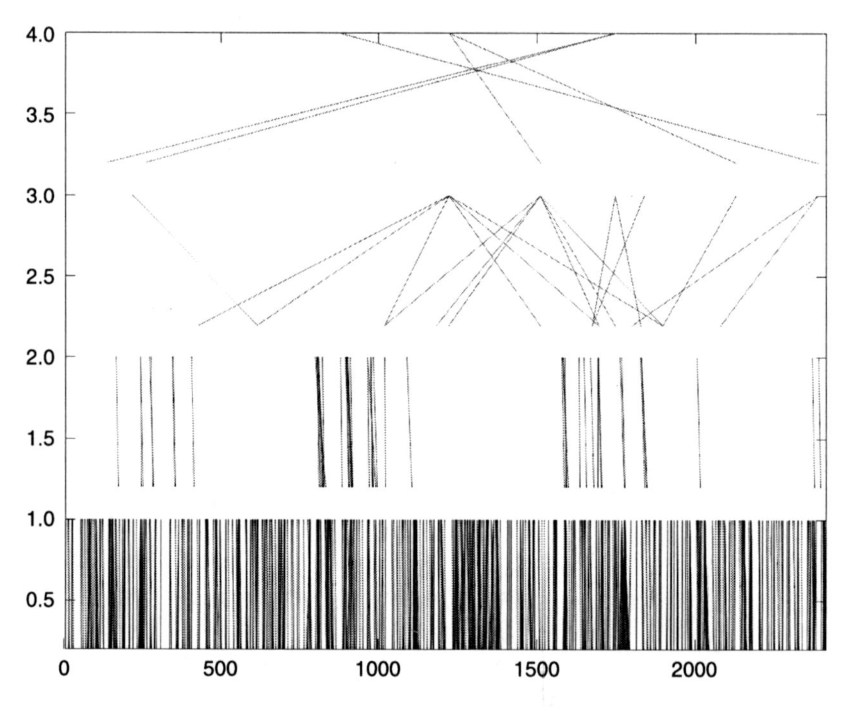

Figure 2.1 A five-layer multiple scale tagging structure of chromosome 22 when $r^2 = 0.4$. The *x*-axis shows SNP indexes sorted by chromosomal positions. The *y*-axis represents the layer number. Chromosome 22 is defined as layer 0. Each layer is constructed iteratively by the tagging process of SNPs of the immediate lower layer. Lines in the graph represent the relationship between taggers in the upper layer and taggees in the lower layer. The distance between taggers and taggees increases as the layer increases, due to increasing window size. (The figure was produced by Octave.)

Tagging is conventionally applied to a chromosomal region, with a fixed window size, due to computational limitations. By extending the concept of tagging from one layer into multiscale tagging, we can achieve the tagging of genome-wide variants. This can be achieved by a bottom-up layer construction approach (Figure 2.1). The entire set of genomic variants is defined as layer 0. We then tagged on top of layer 0, using a defined r^2 threshold and a window size, to produce layer 1 containing only the tagged variants. The tagging reduced the number of variants in

Published by Woodhead Publishing Limited, 2013

layer 1 compared with layer 0. A hierarchical structure can be further constructed, layer by layer iteratively, with an increasing window size. Every layer comprises the tagging SNPs of the immediately lower layer. After a few iterations, the top layer is constructed, serving as a reasonable representation of genome-wide variants.

2.2.3 Haplotypes

The complex disease common variant (CDCV) hypothesis has been the underlying hypothesis of mainstream genome studies in the first decade of the 21st century (Goldstein, 2009). It influences many genome-wide association studies (GWAS) and early phases of the HapMap Project. This hypothesis assumes that complex disease is more likely to be caused by SNPs with higher MAFs (i.e. common alleles) than rarer SNPs and mutations. Therefore, higher priority is given to the searching, genotyping and disease association of common SNPs. But how about the rarer SNPs, mutations and other types of rare variants, which are also likely to be responsible for many diseases? Apart from using next generation sequencing (NGS) to capture all of them, an alternative way to accommodate them is by employing the concept of haplotypes.

The human genome comprises two haploid (i.e. single copy) genomes, one from the father and one from the mother. When we perform conventional genotyping, we usually obtain diploid information, without distinguishing which is from the father and which is from the mother. For example, the "C vs. T" SNP of a person could be one of three diplotypes: CC, CT and TT. A haplotype block is a chromosomal region between two recombination hotspots. Haplotypes are alleles physically residing in a haplotype block of the same haploid chromosome, also called the phase of the genome, as opposed to the "unphased" diplotypes. Since the variants in a haplotype block are assumed to be inherited together, it is possible that a haplotype allele may carry an untyped, disease-causing variant. By analyzing the association between haplotype and the disease trait, we indirectly find the disease-causing variant (Figure 2.2).

A haplotype analysis has many steps:

Determine haplotype blocks

The haplotype block of interest can be identified between two recombination hotspots. This is based on the high LD among the variants (usually common SNPs) within the region.

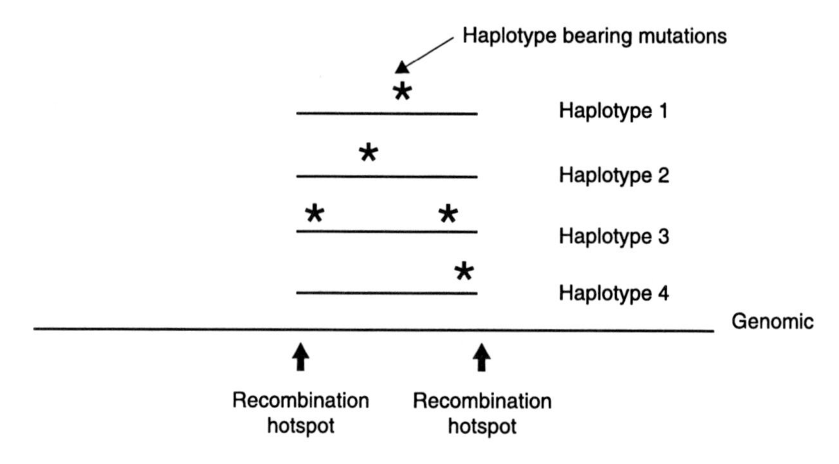

Figure 2.2 The relationship between haplotypes and potential causative variants

Haplotyping

Given a cohort of subjects, the diplotypes of the variants are used for computational phasing (also known as haplotyping) to resolve the phase of the haplotypes. Given a haplotype block of n bi-allelic SNPs within a block, the resolved haplotype number will be much smaller than 2^n, due to the LD between the SNPs. Computational phasing can be carried out by software such as Phase2 or THESIAS (Tregouet and Garelle, 2007) (Box 2.1).

Haplotype association

The haplotype frequency is calculated based on the haplotype counts of the case and control groups. If there is a significant difference between the two groups, then an association is found (Liang and Wu, 2007).

Haplotypes have shown tremendous value in another application: handling mixture genomes. One example is to reconstruct the fetus genome from pregnant women's peripheral blood samples (Kitzman et al., 2012; Fan et al., 2012).

2.3 Genomic platforms and platform level analysis

Genomic studies begin with the acquisition of genomic sequences and their variants, such as point polymorphisms, mutations, copy number

Published by Woodhead Publishing Limited, 2013

alterations and genome rearrangements. The nucleotide bases of DNA and RNA not only contribute to the basic material for genomics and transcriptomics, but also infer theoretical protein sequences, which have a wide use in many aspects of proteomics studies. The NGS platforms and high density SNP arrays are two major high throughput platforms to generate the genomic data.

2.3.1 Capillary Sanger sequencing and next generation sequencing

Conventional capillary sequencers employed the Sanger methodology and served as the major contributors of the human genome project. This method has limitations on sequence reads (~1000 bases), which is far shorter than the genome of most species. Thus, sequencing of genomic DNA needs to start by breaking the DNA macromolecule into smaller pieces. Sequencing is then performed on the pieces. The final step is to computationally or manually stitch the pieces together into a complete genome, an inverse problem called assembly. NGS platforms, so-called because of their relative advancement of conventional capillary sequencers in terms of higher speed and lower cost, have been made commercially available. Major vendors (and platforms) include Illumina (Solexa), Roche (454) and Applied Biosystems (SOlid).

However, NGS produces shorter sequence readings than conventional capillary sequencing. The strength of NGS (speed and lower cost) enables a mass production of readings, resulting in deep sequencing with high coverage (i.e. 100x), which may compensate for the weakness of shorter readings. It has been shown that the sequence readings of commercial NGS platforms can be used for the *de novo* assembly of human genomes (Butler et al., 2008; Gnerre et al., 2011). NGS can also be used to detect germline and somatic point mutations, copy number alterations and structural variations. With the launch of NGS, large-scale genomics investigations now employ NGS in the interest of speed and cost. Conventional capillary sequencing is reserved for small-scale studies. Sequencing technology is still advancing so that new technology (i.e. nanopore) is continuously being developed, so as to reduce the cost and increase the efficiency of sequencing even further.

Base calling for capillary sequencing

The output of capillary sequencers are chromatograms, which are four-channel analog signals representing a digital sequence of nucleotides A,

T, G and C. Chromatograms are commonly stored in trace files. Base calling is the very first step to convert the analog signal into digital nucleotides. A series of software, phred, phrap and consed, has been widely used for handling chromatograms from capillary sequencers. They also serve as the basis of bioinformatics pipelines in many genomic centers (Box 2.1). The phred software is designed to call bases, in the meantime providing a quality score for each base, which is informative for subsequent analysis such as assembly and small-scale variant detection (Ewing and Green, 1998; Ewing et al., 1998). Phrap is a sequence assembler that stitches sequence readings from phred. Consed is a program for viewing and editing phrap assemblies. Software such as Chromas Lite and UGene can be used to visualize the trace files and also produce base calls from them (Box 2.1). The above two software packages do not handle variants (i.e. SNPs in the diploid genome), because the computation is based on the assumption that no heterozygous bases occur in the sequences. Where the variants in sequences are of concern, more advanced software for handling variants should be used.

Base calling for NGS

This is often taken care of by the vendor software, which converts analog signals (image intensities) into digital sequence readings directly. All subsequent analysis relies on these readings.

Variant detection

Detection of small-scale germline or somatic variants can be by the concurrent analysis of multiple samples in the same sequence region. Several high-quality software packages, such as PolyPhred, PolyScan (Chen et al., 2007), SNP detector (Zhang et al., 2005) and NovoSNP (Wecks et al., 2005), are designed for such a purpose (Box 2.1). NovoSNP has an excellent graphical user interface and is suitable for personal use. It can conveniently align multiple trace files so that the candidate variants can be compared across sequences. Polyphred, polyscan and SNP detectors rely on phred for individual base call and quality scores. For example, the integrated pipeline PolyScan is designed to call small insertions and deletions as well as single base mutations and SNPs on multiple traces. It requires not only the called bases but also the quality scores from the phred software and the sequence alignments from consed. The bases are designated again by four features extracted from the trace

Published by Woodhead Publishing Limited, 2013

curves: position, height, sharpness and regularity (Chen et al., 2007). Bayesian statistics are used for the inference of variants.

2.3.2 From sequence fragments to a genome

Assembly

Genome assembly is an important technique to stitch together the sequence readings from sequencers into chromosome-scale sequences. This is how reference genomes (including human) are built. Reference sequences are of tremendous value in all aspects of biology. Once they are built, all the following sequencing tasks of the same species (called re-sequencing) can make use of the reference sequence. The sequence readings can be mapped to the reference using an aligner such as Bowtie (Box 2.1). When a reference genome is unavailable, then direct assembly from sequence readings (called *de novo* assembly) is still required for organizing sequence readings.

Mathematically, a *de novo* assembly of genomic sequences is an inverse problem. The performance of a *de novo* assembly of a genome needs to be gauged by several aspects, including completeness, contiguity, connectivity and accuracy. To understand these, we need first to define "contig" and "scaffold". A contig is a continuous sequence (without gaps) assembled from sequence readings. A scaffold is formed by contigs with gaps of known lengths between the contigs. The contiguity and connectivity can be shown as the average lengths of contigs and scaffolds respectively; the longer, the better. The completeness of contigs and scaffolds is the proportions of the reference genome being covered by contigs and scaffolds (including gaps) respectively. The accuracy means the concordance of bases in the contigs and the reference genome (Gnerre et al., 2011).

DNA fragments to be assembled have several types. The first types are the typical fragments, which are usually shorter than twice the average read lengths. By sequencing the fragments from both ends, an overlapped read region can occur in the middle to join together the two ends, resulting in a single fragment sequence. For example, if the read length is 100 bases, then fragments of 180 bases (shorter than 200 bases) can be unambiguously sequenced. A high volume of fragments can serve as the foundation of assembly of simpler genomes.

Repetitive regions occur in a complex genome such as human and mouse. To overcome the challenge of assembling such genomes, longer fragments of a few thousand bases may be required. This can be done by a mate pair strategy to fuse the two ends of the long fragments (e.g. 3000 bases) to produce a circular DNA, then shatter the DNA into smaller pieces. Since the original two ends are now joined together, they can be sequenced to provide "pairs" of sequences with a longer distance between them. These are called "jump" sequences.

Phrap is one of the pioneering sequence assemblers, which can handle the assembly of readings from capillary sequencers well. As the NGS now produces much shorter readings than conventional capillary sequencing, a new generation of assembler is required. Despite challenging, it has been shown that *de novo* assembly of the repeat rich human genome can be achieved using shotgun NGS reads (Gnerre et al., 2011). The shotgun approach means that genomic DNA is randomly shattered. The assembly of these readings into longer contiguous sequences (called contigs) all relies on computational approaches without knowing of their genomic location beforehand.

ALLPATH-LG is a *de novo* assembly algorithm adapted from the previous ALLPATH algorithm (Butler et al., 2008). ALLPATH-LG was specially tuned for the sequencing of both human and mouse genomes (Box 2.1). It starts by constructing unipaths, which are assembled sequences without ambiguities, then tries to localize the unipaths by joining them together using long "jump" sequences. The human and mouse genome has been shown to be assembled by the same algorithm using the same parameter, suggesting its potential for the genome assembly of other similar species with minimal tuning.

Alignment

Sequence alignment is one of the most extensively discussed bioinformatics topics, which have been the core skill for experimental biologists and professional bioinformaticians alike. It appears in many applications such as the construction of the evolutionary tree or database searches. There are two major types. The "local" sequence alignment aims to find a common partial sequence fragment among two long sequences. The common partial sequences may still have differences in their origins such as insertions, deletions and single-base substitutions. However, the historically earlier "global" sequence alignment is employed to align two sequences of roughly the same size.

The public domain databases, such as NCBI GenBank and EMBL, contain invaluable DNA, RNA and protein sequences of multiple

Published by Woodhead Publishing Limited, 2013

species such as human, rice, mustard, bacteria, fruit fly, yeast, round worm, etc. The sequences are generated by scientists worldwide for many purposes. The NCBI RefSeq database contains curated, high-quality sequences (Pruitt et al., 2012). Public archives often provide many ways to browse through or search for the information contents, and one of the major search methods is by sequence alignment. Finding similar sequences by alignment is of interest, because similar sequences or fragments usually imply similar functions due to their common evolutionary origin. The current model of evolution describes that every organism has originated from a more primitive organism. If a genome duplication event occurs in an ancient organism, then genes in the duplication region will be copied. Each copy of a gene may evolve gradually. It might become a pseudo gene and lose its functionality, or become a new gene with similar functionality. Then these genes are passed through the lineages.

In the past, many algorithms have been proposed for sequence alignments. For example, Needleman–Wursh and Smith–Waterman algorithms are classic examples of global and local sequence alignment respectively. They both employ the dynamic programming approach for optimization. BLAST (Basic Local Alignment Search Tool) is the most widely used method combining a heuristic seed hit and dynamic programming. There are other methods, such as YASS, which employ more degrees of heuristics (Noe and Kucherov, 2005). It is important to know that different algorithms have different characteristics, such as speed and sensitivity.

BLAST is the default search method for the NCBI site. A user can provide a nucleotide sequence of interest by typing in a dialog box, or by submitting a file containing the sequence. After only a few minutes of computation, the system produces a bunch of hits, each of which represents a sequence in the database that has high similarity to the target sequence. If the user clicks on a particular hit, then more details of this sequence will appear.

A variety of indexes are displayed for a particular hit, for example, IR stands for identity ratio, which indicates how much percentage per base is this sequence from the database to the sequence of interest. The e-value stands for expectation value, which is the expected number of coincidence hits given the query sequence and the database.

Many aspects in the system significantly affect the practical usefulness and users' experience in addition to the underlying algorithms. These include visual presentation, scope, completeness and up-to-date information of the database. Certain specialized functionalities can enhance the usefulness greatly. The SNP BLAST site, also provided by NCBI, is such an example. The users still submit the sequences as on the regular BLAST site, but instead of a list of matched sequences, the system reports a list of SNPs and their flanking sequences matched to the submitted sequences. This is particularly useful to identify the location of the submitted sequence in the genome, by means of the high resolution genomic markers. This is also useful for checking the amplicon of the genotyping via sequencing method.

Multiple sequence alignment is used to find the conserved area of a bunch of sequences from the same origin. These sequences are of the same gene family. The conserved area, normally called motifs and domains, is useful in characterizing a gene family.

Sequence alignment can be achieved on-line by using a variety of website services. It can also be done off-line using the downloaded software.

Annotation

When a genome of a species is newly sequenced, it is important to decode the message encoded by the sequence. The sequence structure of a gene is like a language, which can be processed using information technologies. High-quality annotations on the sequences are important to unveil the "meaning" of the sequence's gene prediction. Open reading frame prediction and gene prediction are useful for genome annotation. Some previous works include the use of a hidden Markov model for gene prediction.

2.3.3 Signal pre-processing and base calling on SNP arrays

High-density SNP arrays are the major tool of assessing simultaneously the genome-wide variants and CNVs. Based on intensity profiles, the arrays can be used to detect germline and somatic CNVs, where the unusual copy number (other than the two paternal and maternal copies) is shown as increased or decreased intensity signals. In addition to SNP arrays, array CGH is another major platform to detect the CNVs, particularly the loss of heterozygosity in cancer tissues.

Affymetrix and Illumina are two leading commercial manufacturers and vendors of high density SNP arrays. The latest Affymetrix SNP 6.0 microarray product has 1.8 million probe sets targeting 1.8 million loci evenly scattered in the genome (Table 2.1).

Published by Woodhead Publishing Limited, 2013

Among them, 900,000 are designed to assay SNPs and 900,000 for CNVs. When the number of SNPs is large (e.g. >100), the high-density microarrays have many practical advantages over various conventional single SNP assays. The latter requires a special set of primers for each SNP. This is not only costly and time-consuming, but also complicates the whole experimental process, including the logistics of reagents and management of data. A good bioinformatics system is essential to keep track of data. Finally, unnecessary variability may be introduced when each SNP is handled at different times and by different persons. A high-density microarray can assay thousands of SNPs in one go, reducing the complexity, cost, time and variability significantly.

High-density SNP microarrays contain pairs of probes designed to capture pairs of DNA sequences harboring the major and the minor allele types respectively (i.e. the target sequence). Existence of a particular genotype in the sample will be detected by the signal of hybridization of the probe corresponding to the target sequence. Hybridization will be shown as intensity signals, which are captured by a scanner and stored as digital images. Raw images usually need to be pre-processed, using background subtraction and normalization techniques (for details, see Section 3.3 in Chapter 3), so as to compensate between array variations. The quantitative intensity signal will then be converted to binary showing exist (1) or non-exist (0) of the allele type in the sample, that is, the base calling process. This is a critical process for reading out nucleotide bases from analog signals.

Many unsupervised and supervised algorithms have been proposed for base calling, such as DM, BRLMM, BRLMM2, Birdseed, Chiamo, CRLMM, CRLMM2 and Illuminus. Since the most recent platforms are capable of detecting structural variations together with SNPs, software such as Birdseed can analyze SNP and CNV at the same time.

The handling of sex chromosome (X and Y) data requires special care compared with autosomal chromosomes. Females have two X chromosomes while males have an X and a Y chromosome. Hence, males only have one allele on X (i.e. hemizygous) and the other "missing" allele can be considered as a null allele (Carlson et al., 2006). Since the null allele may affect the base calling performance when a batch of samples contain both males and females, special algorithms have been developed to handle such occurrences (Carlson et al., 2006). Null alleles may also occur in deletion regions of autosomal chromosomes (Carlson et al., 2006).

2.4 Study designs and contrast level analysis of GWAS

2.4.1 Successful factors

Association studies are frequently used to find genes responsible for various clinical manifestations, such as the onset and progression of disease, as well as therapeutic efficacy and toxicity. Association studies were once carried out in a candidate gene fashion, yet now are predominantly done by GWAS. GWAS has successfully identified dozens of novel associations between common SNPs and various common diseases (Hirschhorn, 2009). The knowledge gleaned from these studies is invaluable for the further advancement of biomedical science. Despite some initial success, it was found that most risk alleles have moderate effects and penetrance, preventing them from direct use in personalized healthcare. One of the lessons learned from previous studies is that a well designed study is very important for achieving results. Factors for successful association studies include:

- clinically distinct study groups clearly defined by stringent criteria (deep phenotyping);
- a strong contrast on targeted phenotypes between study groups: the supercase-supercontrol design;
- a negligible contrast on other phenotypes between study groups: no confounding effects;
- no hidden population heterogeneity in the study groups;
- sufficient sample size.

The purpose of all the above factors is to ensure that the detected associations truly represent the clinical trait. A clear and strong contrast in phenotype (i.e. the supercase vs. supercontrol design) could enhance the contrast in genotype, thereby enhancing the possibility of success of finding potential biomarkers, particularly at the exploratory stage. We need to note that the odds ratio (OR) detected does not reflect the true OR in the epidemiological scenario. The hidden population heterogeneity can generally be detected by the inflation factor (Freedman et al., 2004). In such cases, population filters (discussed below) are important.

Many meta-analyses of GWAS have shown that a larger sample size can improve results in terms of both the number of detected SNPs and their significance. For example, a meta-analysis of 46 GWAS with a total

Published by Woodhead Publishing Limited, 2013

of more than 100,000 subjects captured all the 36 originally reported associations, while discovering 59 new associations to personal blood lipid levels (Teslovich et al., 2010). It was found, however, that SNPs on major disease genes of familial hypercholesterolemia were also detected, blurring the conventional categorization of familial diseases vs. common diseases.

2.4.2 The data matrix and quality filters

Association studies are widely used to find the hidden relationship between genetic variants and clinical traits of interest (WTCCC, 2007). In practice, an association study is designed to compare the allele frequencies of two groups with distinct clinical status: the case group comprising subjects with the clinical trait of interest (e.g. a disease state or a therapeutic response) and the corresponding control group. The study subjects are often collected retrospectively, but can also be recruited prospectively. A prospective study means that the clinical outcome is manifested after the samples are collected and assayed (Patterson et al., 2010). The underlying concept for an association study is that genotypes which cause a clinical trait, directly or indirectly, will be more enriched in the case groups, producing a difference in allele frequencies.

In the past, association studies were done by the candidate-gene approach that scrutinizes a targeted set of genes, which is hypothesized to exert effects on disease etiologies based on prior knowledge. This approach has gradually been replaced by GWAS, which is done in a holistic perspective and hypothesis free style with all known genes screened in an unbiased fashion. This was enabled only recently by the maturation of commercial high density SNP assaying platforms such as Affymetrix SNP 6.0 arrays. These platforms were designed for assaying SNPs with not-so-rare minor alleles (e.g. MAF >0.1), a consequence of the "complex diseases common variant" hypothesis widely accepted before 2009 (Goldstein, 2009). A GWAS usually screens 100,000 to 1,000,000 SNPs, a reasonable sub-sampling of the entire set of genomic variants (Table 2.1).

A conventional master file of an association study is a subject-by-SNP data matrix, comprising genotypes in multiple SNPs of multiple subjects. The genotypes of all subjects are converted from the scanned intensity of SNP microarrays or from the sequencing trace files using the corresponding base calling software. The subjects are sorted by clinical status for ease of comparison. The SNPs are preferably sorted by the chromosomal locations. The subject-by-SNP data matrix is denoted as D, where the subjects are categorized as cases or controls. Each row of D represents a sample of a subject (person), and each column represents an SNP. Denote n as the total number of SNPs, therefore, $D = \{SNPj \mid 0 <= j < n\}$. The subject-by-SNP matrix is part of a PLINK format. It can be visualized using colors, as in the SeattleSNPs presentation. In some data formats. such as the Oxford format, FastTagger and Slide, the SNP-by-subject is used instead. Considering the 100,000 to 1,000,000 SNPs screened in a GWAS, the dimension of the dataset appears to be huge. Yet, the underlying correlation between adjacent SNPs is a biological feature known as LD, which can effectively reduce the complexity of the data matrix.

A collection of quality filters is usually applied to the data matrix, prior to or after the statistical analysis, to ensure the results are derived from high-quality data and reflect the genuine biological effects. These filters are usually imposed in the following order:

1. Subject filters:

   a) *Population filter*: remove subjects (persons) from the dataset who belong to a population other than the study population. The population of a subject can be known from self-reported information, or by the analysis of genome-wide genotypes.

   b) *Sample or microarray quality filter*: remove samples (microarrays) showing abnormal intensity distribution.

   c) *By call rate per subject*: the number of called SNPs must be greater than a threshold such as 95% (Tanaka et al., 2009).

2. SNP filters:

   a) *Base calling quality*: remove SNPs that demonstrate abnormal intensity distributions, which render low-quality scores detected by the base calling algorithm.

   b) *Call rate per SNP*: number of subjects which were successfully called in this SNP needs to be greater than a threshold such as 95% (Tanaka et al., 2009). This is particularly critical in the sense that problematic SNPs usually give a strong but false significance if not filtered previously. The problematic SNPs usually have abnormal signal distributions, which are difficult for base calling. That is why the call rate is usually low.

Published by Woodhead Publishing Limited, 2013

c) *Hardy–Weinberg equilibrium test in the control group (optional for case group)*: SNPs that deviate from the Hardy–Weinberg equilibrium are removed. The Hardy–Weinberg equilibrium means that the observed genotype counts are close to the expected genotype counts calculated from observed allele frequencies. Whether the Hardy–Weinberg equilibrium is deviated from can be calculated using Chi-square tests. This is usually applied to the control (healthy subjects) group only (Minelli et al., 2008).

d) *Association of nearby SNPs (optional)*: This is done after the calculation of association. This is based on observations that the risk locus usually manifests itself as multiple adjacent hits due to their LD (WTCCC, 2007). If the association is a singleton, which may be due to some artifacts rather than real biological effects, they are thus removed.

The data matrix could therefore contain missing data (blanks) due to the limitation of the base calling algorithm (the null calls) or the sample quality (filtered out by the quality filters). A small degree of missing data will be tolerated and will not hamper the analysis. Nevertheless, we can also choose to employ imputation algorithms to fill in the missing data, or un-genotyped SNPs, based on the LD structure of the SNPs. Imputation algorithms are particularly useful on the meta-analysis of GWAS employing different platforms, for example, Affymetrix and Illumina arrays, where the assayed SNPs are not identical and need to be imputed.

2.4.3 Contrasts of genotypes of individual SNPs

SNPs manifesting strong contrasts of allelic or genotypic frequencies between the clinical groups are pursued in an associations study. Two important gauges for every candidate association are:

1. how unlikely an association is false positive;
2. how strong the genotypic contrasts are.

The first gauge is the significance level, also known as the P-values, of statistical tests. The P-values quantify the probability of rejecting the null hypothesis when there is no association; that is, the probability of a false positive. Hence, the smaller the P-value, the higher the statistical significance. SNPs with P-values smaller than a predefined threshold

Table 2.2 The 2 × 2 contingency table of allele counts for an allelic association

		Clinical Status		
		Control	Case	Sum
Alleles	Allele 1	C1	C2	C1 + C2
	Allele 2	C3	C4	C3 + C4
	Sum	C1 + C3	C2 + C4	C1 + C2 + C3 + C4

Table 2.3 The 3 × 2 contingency table of genotype counts

	Clinical Status		
	Control	Case	Sum
Genotype 1 (Alleles 1 & 1)	C1	C2	C1 + C2
2 (Alleles 1 & 2)	C3	C4	C3 + C4
3 (Alleles 2 & 2)	C5	C6	C5 + C6

(commonly set at 0.05 when only one variant is assessed) are confidently associated to a particular trait of interest. The second gauge is the ORs.

Both P-values and ORs are calculated from the counts of different allelic and genotypic forms. The 2 × 2 and 2 × 3 contingency tables are usually employed to present the counts (Tables 2.2 and 2.3).

Association tests include the allelic, genotypic and trend tests, which can be formulated either as a Chi-square test or a Fisher's exact test. These tests are used to detect three types of risk patterns, the additive, dominant and recessive modes. Allelic tests evaluate the difference of allelic frequencies between the case and control groups. A related trend test aims to examine the increasing or decreasing trends of diseased frequencies (i.e. proportion of subjects who are the case group) in response to the addition of risk alleles. Allelic and trend tests basically evaluate the additive effect of each risk allele, in the sense that genotype 3 of two risk alleles will have twice the effect as genotype 2 with only one risk allele, thus the relationship between alleles and disease frequency is approximately linear. However, genotypic tests evaluate the difference of genotype frequencies between the case and control groups. Genotypic tests have three variants:

Published by Woodhead Publishing Limited, 2013

1. a comparison of three genotypes;
3. grouping genotypes 1 and 2 together, to be compared with genotype 3 (representing a recessive mode of inheritance);
4. grouping genotypes 2 and 3 together, to compare with genotype 1 (representing a dominant mode of inheritance).

Since the human genome is diploid, an unphased genotype (i.e. the diplotype) contains two alleles. Hence, the total allelic count is twice as large as the total genotype counts. This makes the allelic test double the sample size than the genotypic tests.

For example, the mathematical equation for a Chi-square allelic is: the Chi-square value (X^2)

$$X^2. = \sum_{k=1}^{4} \frac{(C_k - E_k)^2}{E_k}$$

where

$$E1 = (C1 + C2)^*(C1 + C3)/(C1 + C2 + C3 + C4)$$

$$E2 = (C1 + C2)^*(C2 + C4)/(C1 + C2 + C3 + C4)$$

$$E3 = (C1 + C3)^*(C3 + C4)/(C1 + C2 + C3 + C4)$$

$$E4 = (C2 + C4)^*(C3 + C4)/(C1 + C2 + C3 + C4)$$

and the degrees of freedom is 1.

ORs are the ratio of odds with and without a particular allele or genotype. They can be formulated as:

$$OR = (Pr(D|G) / Pr(\sim D|G)) / (Pr(D|\sim G) / Pr(\sim D|\sim G))$$

Under the same formula, OR has many meanings, depending on the definition of G and ~G. An odds ratio could mean:

- odds ratio per allele;
- odds ratio of the recessive mode;
- odds ratio of the dominance mode;
- heterozygote odds ratio;
- homozygote OR.

The corresponding definitions of G and ~G are

- G is an allele, while ~G is the other allele;
- G is an genotype, while ~G represent the other two genotypes;

- G represents two genotypes, while ~G is the other genotype;
- G is genotype 2, while ~G is genotype 1;
- G is genotype 3, while ~G is genotype 1.

It is thus important to check the definition before interpreting a result from the literature, particularly when multiple research works are compared. For example, the ORs in Tanaka et al. (2009) are the dominance mode, while those in WTCCC (2007) are heterozygote and homozygote ORs.

A similar concept is relative risk (RR), defined as:

$$(\Pr(D|G)) / (\Pr(D|\sim G))$$

2.4.4 Sample size and multiple comparisons

When genome-wide variants are evaluated in parallel, multiple hypotheses testings, one per a variant, are performed. This comes with the cost that we must tighten the stringency on P-values for each hypothesis. This is called the issue of multiple comparisons. In such types of study, we need to control the family-wise type 1 error rate of the entire study. A common way to address the family-wise error rate is to employ Bonforroni corrections. For example, if n SNPs are assayed in a GWAS, and we want to control the family-wise error rate below 0.05, then each SNP will have an individual P-value smaller than 0.05/n before it can be declared statistically significant. For an Affymetrix SNP6.0 microarray with 900,000 genome-wide SNPs, significance level per SNP based on Bonforroni correction is $5*10^{-8}$.

In the Bonforroni correction setting, n represents the number of independent tests. However, due to the LD between adjacent SNPs in the human genome, hypothesis testings on these SNPs are not independent. Therefore, the conventional Bonforroni correction is often too conservative and inadequate. This can be mitigated by two approaches.

First, a permutation test is a non-parametric alternative that can take LD into consideration. It uses empirical permutation to calculate family-wise error rates, therefore requires a heavy computation which needs to be overcome, particularly in GWAS settings. Permory is a good software of permutation tests on GWAS datasets (Pahl and Schafer, 2010).

Published by Woodhead Publishing Limited, 2013

Second, prior domain knowledge can be used to devise filters and integrators (Ideker et al., 2011) and reduce the number of independent tests. Filters are used to remove poor-quality data and variants unrelated to the study. Integrators can be used to group variants into higher-order entities such as genes or pathways.

The sample size of this study depends on five factors. The first is the expected effect size of associations, often quantified by ORs. The second issue is the MAF. The range of MAF of SNPs is between 0.01 and 0.5. The third factor is the significance level considering the multiple comparison issues. The fourth factor is statistical power, which is usually set at 80%. The fifth factor is the ratio of the two clinical groups to be compared.

2.4.5 Visual presentation and interpretation

A contrast level analysis of GWAS data will derive a series of P-values for hundreds of thousands of individual SNPs scattered in all chromosomes. A Manhattan plot is a scatter plot that offers a useful grand-scale visualization of the GWAS result. The assayed SNPs are sorted by chromosome locations in the x-axis. The y-axis shows the negative logarithm of P-values, and as a result the higher dots represent significant hits. Freeware, such as WGAviewer (Ge et al., 2008) and Goldsurfer2 (Pettersson et al., 2008), as well as commercial software such as Golden Helix, can be used to produce a Manhattan plot (Figure 2.3).

GWAS and its subsequent validation render a short-list of SNPs, which are statistically associated to clinical traits. Albeit intriguing, the molecular mechanisms behind the associations are still vague. Thus, it is essential to pursue further functional exploration of detected associations. The first step is an *in silico* check of the SNP (variant) location in relation to the genes and the local LD pattern. Genome browsers in the RefSNP, dbSNP or HapMap websites can help users to locate whether the SNP is in the exon, intron or intergenic regions. If the SNP is located in the intron, then the databases H-DBAS or EBI-ASTD can be used to scrutinize the alternative splicing patterns. H-DBAS can also illustrate whether the SNP is in coding or UTR regions. It also offers a nice comparison with the mouse transcripts.

A check of the functional domain where the SNP resides requires a protein-level database, such as Pfam or InterPro (Box 2.1). A complete understanding of the association requires data from further experiments on cell or tissue specific gene expressions.

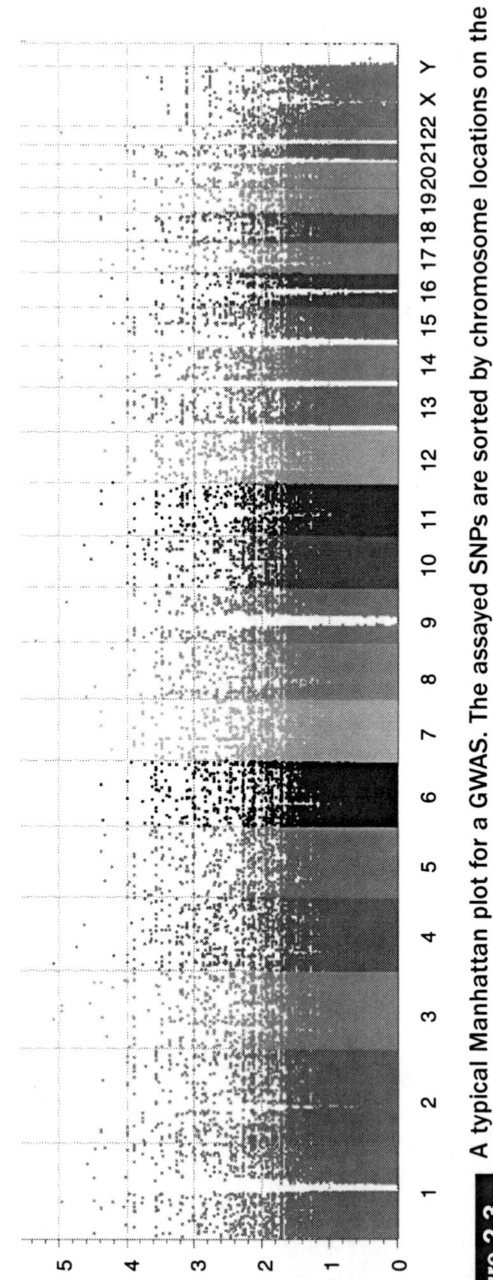

Figure 2.3 A typical Manhattan plot for a GWAS. The assayed SNPs are sorted by chromosome locations on the x-axis, with different chromosomes showing different colours. The y-axis shows the negative logarithm of p-values; as a result the higher dots represent significant hits. (The plot was produced by Golden Helix.)

2.4.6 The missing heritability

The GWAS approach has successfully identified and validated hundreds of disease prone (and occasionally disease bearing) SNPs. These SNPs may serve as clinically useful biomarkers and biosignatures for disease susceptibility and prognosis, as well as revealing disease mechanisms (Hirschorn, 2009). However, this approach also has its limitations. It has been observed that GWAS-detected alleles are of low penetrance and that the total variability explained by all these SNPs together is only a fraction of the variation caused by heritability (Doucleff, 2010; Donnelly, 2008; Goldstein, 2009; McCarthy, 2008, Manolio et al., 2009). It is thus suggested that personal disease risks are not only contributed by common SNP alone, but also by other forms of genomic variations such as CNVs or rare mutations, as well as their synergistic effects (Goldstein, 2009; McClellan and King, 2010; Manolio et al., 2009).

Among these possible sources of missing heritability, the accumulation of multiple mutations (i.e. MAF <1%) for disease etiology deserves most attention. Lessons from the familial diseases, such as familial hypercholesterolemia, familial breast cancer, hereditary nonpolyposis colorectal cancers and cystic fibrosis, tell us that numerous rare germline mutations play significant roles. From an evolutionary perspective, all SNPs originate from mutations. They are actually two sides of the same coin (McClellan and King, 2010). The importance of mutations justifies the sequencing technology as a critical tool for detecting the genomic etiology of disease.

2.5 Adaptive exploration of interactions of multiple genes

Complex diseases are the result of multiple abnormal genes in multiple pathways (Phillips, 2008). A multiple hit etiology (causation) is commonly ascribed to complex diseases such as autoimmune diseases and cancer (Goodnow, 2007). In the multiple hit model, the effect of individual risk factors is mild and obscure. Only a combination of them will trigger the disease. Conventionally, GWAS data were analyzed mostly on the single SNP basis (Cordell, 2009). Current data of GWAS show that individual variants often do not exhibit large enough differences (effect size) to stratify patients. Thus, multivariate analysis on SNP combinations is a critical augmentation to the analysis on the individual variant basis.

One foreseeable challenge for the analysis of SNP combinations is the enormous search space (Cordell, 2009). Current GWAS usually involves hundreds of thousands of SNPs. Potential pairwise combinations among them are already numerous, let alone higher-order combinations. An exhaustive search of higher-order combinations is almost unfeasible. It has been proposed to restrict the search space using prior information such as pathways or protein–protein interaction networks. However, these strategies cannot identify *de novo* combinations, which are particularly attractive under current incomplete knowledge.

An adaptive algorithm is thus introduced here for the exploration of a large search space, using iterative trial and error to adapt itself to better solutions. The implementation of this adaptive algorithm is by way of genetic algorithms and Boolean algebra, the latter of which is a bivalent algebraic system consisting of 0 and 1 (Liang et al., 2006; Hsieh et al., 2007). Boolean algebra includes addition (+; AND), multiplication (·; OR) and negation (–), corresponding to union, intersection and complement in the set theory, respectively.

An adaptive model, denoted as M, classifies subjects as one of two phenotypes coded by values of 0 or 1. The value of M, determined by a combination of model elements (mi), also coded by values of 0 and 1, is joined together by a series of Boolean operators, "AND" and "OR". An "OR" logic associates two distinct etiologies for the same phenotype; an "AND" operator represents a two-hit causation. Using a biallelic A and T SNP as an example, a model element is associated to a genotype in either the recessive mode:

mi: SNPk = "AA"

or a dominant mode of inheritance:

mi: SNPk = "AA" or "AT".

The genetic algorithm is a modern heuristic method for solving combinatorial optimization tasks. The task of model construction aims to maximize the fitness score, which is defined as the classification performance of the model M of the entire cohort (see Section 6.2 in Chapter 6 for performance indexes). A random model generator is required to initiate the computation. First, the number of model elements is randomly determined. Then, a series of variants are randomly chosen for all model elements. Each SNP vs. model element relationship has four possible types for dominant and recessive modes of inheritance, such as "AA", "AA or AT", "TT" or "AT or TT". The additive (+) and

Published by Woodhead Publishing Limited, 2013

multiplicative (·) Boolean operators are then randomly chosen between the model elements. Finally, a negation (–) operator is randomly determined whether or not to appear in front of the entire statement. The algorithm employs mutation and crossover operations for altering an existing model.

Five different types of mutations are employed in the algorithm:

1. element insertion;
2. element deletion;
3. element substitution;
4. operators ·/+ swap; and
5. case and control swap.

The element insertion operation introduces a new random element into the model, increasing the model length by 1. The element deletion operation removes an element from the model. The element substitution operation changes the specified genotypes in a model element, for example, from SNPk = "AA" to "AA or AT". The operators ·/+ swap converts a multiplication (·) into an addition (+), or vice versa. This operation changes the nonlinear relationship between the model elements. For example, if this operation modifies the model $M = m1 + m2 \cdot m3 \cdot m4$ as $m1 + m2 + m3 \cdot m4$, then the relationship between the elements is changed. Finally, the case and control swap introduces a negation operator in front of the model. If there is already a negation operator, then this operation effectively removes the original negation operator.

The crossover operation is analogous to the chromosomal recombination events occurring in meioses of cell cycles. The rationale for the crossover operation is that if the good performance of two models is mainly due to parts of themselves, then a crossover operation may combine these two parts, resulting in a scrutiny in the proximity of the search space in the previous two models.

Using the defined operations, the models with higher fitness scores are randomly mutated and crossed over with each another so as to produce various candidate models, exploring the entire solution space in a systematic manner. Each of these models is used to predict the samples in the training dataset. The prediction performances of the models are then evaluated by their fitness scores. Models and their elements with higher fitness scores are preserved and also serve as templates for constructing the models in the next iteration. The same concept has been used to optimize the classification models by multiple haplotype (Liang and Wu, 2007).

2.6 Somatic genomic alterations and cancer

Cancer is thought to be a genetic disease where a collection of somatic mutations drives its development (Hanahan and Weinberg, 2011). Chromosomal instability is a prominent characteristics of cancer cells. Cancer cells are alive and evolving constantly. It has long been hypothesized that cancer cells go through a mini-evolution process in the body and that it continues to acquire unusual capability by a series of somatic mutations of various scales. In other words, cancers are driven by chromosomal mutations. With the ample sample source, and the availability of high-throughput sequencing methods to detect mutations, we can now explore this in a more complete scale. Sequencing-based studies can detect somatic mutations directly, rather than surrogate signals as in GWAS studies. But sequencing-based methods also introduce new challenges. Many mutations are found in the cancer genome. Some can drive the disease, while others may be neutral. The neutral mutations are often called passive mutations (Ashworth et al., 2011). It is difficult to prioritize the mutations to confirm which mutation actually “drives” the progression of disease.

To provide evidence and details of this theory, two recent studies have investigated somatic point mutations on tumor cells by NGS. By comparing the somatic mutations on the primary site and metastatic site of pancreatic cancer tissues, a detailed lineage of cancer cells is revealed by their mutation patterns (Yachida et al., 2010; Lueberk, 2010). The primary site already develops many clones, which are equipped with necessary mutations for the success of developing metastasis in other organs. Cancer is thus shown to be a progressive disease and the entire course can take as long as 15 years.

Increase of copy numbers of oncogenes and decrease of copy numbers of tumor suppressor genes may drive the progression of cancer. To investigate the alterations of copy numbers in the cancer genomes, a recent study investigated 3131 tumor samples with 26 cancer types using high-density SNP microarrays (Beroukhim et al., 2010). It was found that tumor suppressor genes such as RB1 and CDKN2A/B are frequently deleted, while oncogenes such as EGFR, FGFR1 and MYC were frequently amplified. The frequently altered regions in the tumor specimens are called somatic copy number alterations (SCNA), to differentiate them from germline CNVs. Furthermore, oligonucleotide array comparative genomic hybridization technology has been used to

Published by Woodhead Publishing Limited, 2013

detect the frequent deletions of immunoglobulin heavy chain and T cell receptor gene regions in chronic myeloid leukemia using PBMC samples (Nacheva et al., 2010).

2.7 Case studies

2.7.1 Chromothripsis on cancer progression

One major characteristic in the cancer genome is its pervasive structural variations caused by extensive rearrangements. It has long been hypothesized that the structural variations gradually accumulate over time. Contrary to this model, a mechanism called chromothripsis was recently discovered. It refers to localized massive rearrangements occurring in one or few chromosomes, which may occur in one catastrophic event (Stephens et al., 2011).

Stephens et al. (2011) observed an infrequent, somatic, localized rearrangement pattern of the cancer genome based on their NGS and SNP microarray assays. This pattern has three major features:

1. copy number oscillates between 1 and 2 in a localized region;
2. the breakpoints are clustered and reused multiple times;
3. the co-joined segments were not adjacent originally.

They hypothesized that this pattern is caused by rearrangements occurring in a single catastrophic event, rather than a series of events. They employed a Monte Carlo method to simulate both the one-event and the progressive models, and concluded that the one-event model better depicts their observation. Furthermore, this event is shown to occur early, due to the paucity of other mutations occurring alongside the chromothripsis. Thus, this early event, although infrequent, disrupts multiple genes so as to drive the further progression of cancer. This example demonstrates the analysis of the cancer genome to shed light on an obscure yet important molecular event during the progression of the disease.

2.7.2 Ancestry inference and population analysis based on genomic variants

High resolution genomic variants offer rich information about human population and ancestry, in addition to all the health-related messages mentioned above. A few personal genome service companies, such as decodeMe, 23andMe and Navigenics, have emerged since 2005 upon the availability of technology. All of them provide population analysis for individual customers. Conventionally, ancestry inference and population analysis were carried out using mitochondrial DNA or the Y chromosome. Now with the availability of genome-wide and high resolution variant information, a higher resolution on ancestry inference and population analysis can certainly be achieved.

Li and Durbin (2011) demonstrate the use of genome-wide variants for the inference of human ancestry and population history. The analysis is based on the genome of seven subjects. The local density of heterozygosity is the key clue to the estimation of local time.

Behar et al. (2010) provide a population analysis of Jewish people based on genotypes of 121 persons from Illumina 610 K and 660 K bead array. They used principal component analysis (PCA) to project the samples in a 2-D plane spanned by two major principal components, which correspond nicely to two geographical axes. The technique of PCA will be introduced in Chapter 3.

2.8 Take home messages

- Reference genomes are constructed by the *de novo* assembly of sequence reads.
- Sequence alignments are basic skills to find homologous sequences from the public domain database.
- A spectrum of variants (SNP, CNV and structural variations) exist in the human genome.
- Association studies aim to associate the variant genotypes to clinical traits, thereby revealing phenotype critical genes.
- The strength of association between genomic variants and clinical traits is quantified by P-values and ORs.
- Manhattan plots can be used for genome-wide presentation of association.
- Adaptive models are useful for capturing clinical genetic signatures.
- Somatic variants, particular the structure variations, are responsible for cancer initiation and progression.

Published by Woodhead Publishing Limited, 2013

2.9 References

Altshuler, D.M., Gibbs, R.A., Peltonen, L., et al. (2010) Integrating common and rare genetic variation in diverse human populations. *Nature*, **467 (7311)**: 52–8.

Ashworth A., Lord, C.J. and Reis-Filho, J.S. (2011) Genetic interactions in cancer progression and treatment. *Cell*, **145(1)**: 30–38.

Behar, D.M., et al. (2010) The genome-wide structure of the Jewish people. *Nature*, **466**: 238–42.

Beroukhim, R., Mermel, C.H., Porter, D., Wei, G., Raychaudhuri, S., et al. (2010) The landscape of somatic copy-number alteration across human cancers. *Nature*, **463**: 899–905.

Butler, J., MacCallum, I., Kleber M., et al. (2008) ALLPATHS: *de novo* assembly of whole-genome shotgun microreads. *Genome Res.* **18(5)**: 810–20.

Carlson, C.S., et al. (2006) Direct detection of null alleles in SNP genotyping data. *Hum. Mol. Genet.* **15(12)**: 1931–7.

Carlson, C.S., Eberle, M.A., Rieder, M.J., Yi, Q. and Kruglyak, L. (2004) Selecting a maximally informative set of single-nucleotide polymorphisms for association analyses using linkage disequilibrium. *Am. J. Hum. Genet.*, **74**: 106–20.

Chen, E.Y., et al. (2007) PolyScan: An automatic indel and SNP detection approach to the analysis of human resequencing data. *Genome Res.*, **17**: 659–66.

Cordell, H.J. (2009) Detecting gene-gene interactions that underlie human disease. *Nat. Rev. Genet.*, **10**: 392–404.

DeCerbo, J. and Carmichael, G.G. (2005) SINEs point to abundant editing in the human genome. *Genome Biol.*, **6(4)**: 216–19.

Donnelly, P. (2008) Progress and challenges in genome-wide association studies in humans. *Nature*, 456: 728–31.

Doucleff, M. (2010) Genomics select. *Cell*, **142(2)**: 177.

Ewing, B. and Green, P. (1998) Base-calling of automated sequencer traces using phred. II: Error probabilities. *Genome Res.*, **8**: 186–94.

Ewing, B., Hillier, L., Wendl, M.C. and Green, P. (1998) Base-calling of automated sequencer traces using phred. I: Accuracy assessment. *Genome Res.*, **8**: 175–85.

Fan, H.C., Gu, W. and Wang, J., et al. (2012) Non-invasive prenatal measurement of the fetal genome. *Nature*, 487(7407): 320–4.

Feuk, L., Carson, A.R. and Scherer, S.W. (2006) Structural variation in the human genome. *Nat. Rev. Genet.*, 7: 85–97.

Freedman, M.L., Reich, D., Penney, K.L., McDonald, G.J., Mignault, A.A., et al. (2004) Assessing the impact of population stratification on genetic association studies. *Nat Genet.*, **36(4)**: 388–93.

Ge, D., Zhang, D., Need, A.C., Martin, O., Fellay, J., Telenti, A. and Goldstein, D.B. (2008) WGAViewer: Software for genomic annotation of whole genome association studies. *Genome Res.*, 18(4): 640–3.

Gnerre, S., et al. (2011) High-quality draft assemblies of mammalian genomes from massively parallel sequence data. *PNAS*, **108**: 1513–1518.

Goldstein, D.B. (2009) Common genetic variation and human traits. *N. Engl. J. Med.*, **360**: 1696–8

Goodnow, C.C. (2007) Multistep pathogenesis of autoimmune disease. *Cell*, **130**: 25–35.

Gupta, G.P. and Massague, J. (2006) Cancer metastasis: Building a framework. *Cell*, **127**: 679–95.

Hanahan, D. and Weinberg, R.A. (2011) Hallmarks of cancer: The next generation. *Cell*, **144**: 646–74.

Hirschhorn, J.N. (2009) Genome-wide association studies: Illuminating biologic pathways. *N. Engl. J. Med.*, **260**: 1699–701.

Hsieh C.H., Liang K.H., Hung, Y-J., Huang, L-C., Pei, D., et al. (2006) Analysis of epistasis for diabetic nephropathy among Type 2 diabetic patients. *Hum. Mole. Gen.*, **15(18)**: 2701–2.

Ideker, T., Dutkowski, J. and Leroy Hood, L. (2011) Boosting signal-to-noise in complex biology: Prior knowledge is power. *Cell*, **144**: 860–3.

IHGSC (2004) Finishing the euchromatic sequence of the human genome. *Nature*, **431**: 931–45.

Kitzman, J.O., Snyder, M.W., Ventura, M., et al. (2012) Non-invasive whole-genome sequencing of a human fetus. *Sci Transl Med.*, **4(137)**: 137–76.

Lander, E.S., Linton, L.M., Birren, B., et al. (2001) Initial sequencing and analysis of the human genome. *Nature*, **409**: 860–921.

Langmead, B. and Salzberg, S. (2012) Fast gapped-read alignment with Bowtie 2. *Nature Meth.*, **9**: 357–9.

Langmead, B., Trapnell, C., Pop, M. and Salzberg, S.L. (2009) Ultrafast and memory-efficient alignment of short DNA sequences to the human genome. *Genome Biol.*, **10**: R25.

Li, H, and Durbin, R. (2011) Inference of human population history from individual whole-genome sequences. *Nature*, **475(7357)**: 493–6.

Liang, K.H. and Wu, Y-J. (2007) Prediction of complex traits based on the epistasis of multiple haplotypes. *J. Hum. Gen.*, **52(5)**: 456–63.

Liang, K.H., Hwang, Y., Shao, W-C. and Chen, W.Y. (2006) An algorithm for model construction and its applications to pharmacogenomic studies. *J. Hum. Gen.*, **51**: 751–9.

Lin, L., Shen, S., Tye, A., et al. (2008) Diverse splicing patterns of exonized Alu elements in human tissues. *PLoS Genet.*, **4**: e1000225.

Liu, G., Wang, Y., Wong L. (2010) FastTagger: An efficient algorithm for genome-wide tag SNP selection using multi-marker linkage disequilibrium. *BMC Bioinformatics*, **11**: 66.

Luebeck, E.G. (2010) Genomic evolution of metastasis. *Nature*, **467**: 1053–4.

Manolio, T.A., et al. (2009) Finding the missing heritability of complex diseases. *Nature*, **461**: 747–53.

McCarthy, M.I. (2008) Casting a wider net for diabetes susceptibility genes. *Nat. Genet.*, **40**: 1039–40.

McClellan, J. and King, M-C. (2010) Genetic heterogeneity in human disease. *Cell*, **141(2)**: 210–17.

Minelli, C., Thompson, J.R., Abrams, K.R., et al. (2008) How should we use information about HWE in the meta-analysis of genetic association studies? *Int. J. Epidemiol.*, **37**: 136–46.

Published by Woodhead Publishing Limited, 2013

Muotri, A.R., et al. (2007) The necessary junk: New functions for transposable elements. *Hum. Mol. Genet.*, 16: R159–67.

Nacheva, E.P., Brazma, D. and Virgili, A. (2010) Deletions of immunoglobulin heavy chain and T cell receptor gene regions are uniquely associated with lymphoid blast transformation of chronic myeloid leukemia. *BMC Genomics*, **11**: 41.

Noe, L. and Kucherov, G. (2005) YASS: Enhancing the sensitivity of DNA similarity search, *Nucleic Acids Res.*, **33(2)**: W540–3.

Pahl, R. and Schafer, H (2010) PERMORY: an LD-exploiting permutation test algorithm for powerful genome-wide association testing. *Bioinformatics*, 26(17): 2093–100.

Patterson, S.D., et al. (2010) Prospective-retrospective biomarker analysis for regulatory consideration: White paper from the industry pharmacogenomics working group. *Pharmacogenomics*, **12(7)**: 939–51.

Pettersson, F., Morris, A.P., Barnes, M.R. and Cardon, L.R. (2008) Goldsurfer2 (Gs2): A comprehensive tool for the analysis and visualization of genome wide association studies. *BMC Bioinformatics*, 9: 138.

Pennisi, E. (2010) 1000 Genomes project gives new map of genetic diversity. *Science*, 330: 574–5.

Phillips, P.C. (2008) Epistasis: The essential role of gene interactions in the structure and evolution of genetic systems. *Nat. Rev. Genet.*, 9(11): 855–67.

Pruim, R.J., Welch, R.P., Sanna, S., et al. (2010) LocusZoom: Regional visualization of genome-wide association scan results. *Bioinformatics*, 26(18): 2336–7.

Pruitt, K.D., Tatusova, T., Brown, G.R. and Maglott, D.R. (2012) NCBI Reference Sequences (RefSeq): Current status, new features and genome annotation policy. *Nucleic Acids Res.*, 40(Database issue): D130–5.

Stephens, P.J., et al. (2011) Massive genomic rearrangement acquired in a single catastrophic event during cancer development. *Cell*, 144: 27–40.

Tanaka, Y., Nishida, N., Sugiyama, M., et al. (2009) Genome-wide association of IL28B with response to pegylated interferon- and ribavirin therapy for chronic hepatitis C. *Nature Genet.*, **41(10)**: 1105–9.

Teslovich, T.M., Musunuru, K., Smith, A.V., et al. (2010) Biological, clinical and population relevance of 95 loci for blood lipids. *Nature*, 466(7307): 707–13.

The International HapMap Consortium (2007) A second generation human haplotype map of over 3.1 million SNPS. *Nature*, **449**: 851–62.

Tregouet, D.A. and Garelle, V (2007) A new JAVA interface implementation of THESIAS: Testing haplotype effects in association studies. *Bioinformatics*, **23(8)**: 1038–9.

Wang, K., et al. (2007) PennCNV: An integrated hidden Markov model designed for high-resolution copy number variation detection in whole-genome SNP genotyping data. *Genome Res.*, 17: 1665–74.

Wecks, S., Del-Favero, J., Rademakers, R., et al. (2005) novoSNP, a novel computational tool for sequence variation discovery. *Genome Res.*, 15(3): 436–42.

WTCCC (2007) Genome-wide association study of 14,000 cases of 7 common diseases and 3000 shared controls. *Nature*, **447**: 661–78.

Yachida, S., Jones, S., Bozic, I., et al. (2010) Distant metastasis occurs late during the genetic evolution of pancreatic cancer. *Nature*, **467(7319)**: 1114–17.

Zhang, J., et al. (2005) SNPdetector: A software tool for sensitive and accurate SNP detection. *PLoS Computational Biology*, October.

Published by Woodhead Publishing Limited, 2013

3

Transcriptomics

DOI: 10.1533/9781908818232.49

Abstract: Ribonucleic acids (RNA) are macromolecules with diverse cellular and biological functions, composed of linear chains of nucleotides. RNAs either serve as templates for protein synthesis, or play critical catalytic and regulatory roles. The abundance and activities of all RNAs in time and space, referred to as the transcriptome, offer a global perspective on molecular activity in cells, which jointly affect human physiology and pathology. Microarrays and next-generation sequencing are two major technological platforms for transcriptome studies. An integrative, systems level approach is introduced, aiming to establish the genetic cause of disease by analyzing RNA and DNA profiles concurrently using advanced computational models. The case studies cover important biomedical topics such as immune activation which undermines kidney transplantation, as well as the intricate connection between stem cells and cancer. They together show how novel insights can be revealed by multiple steps of analysis.

Key words: visual presentation, RNA world, gene expression, secondary structure, integrative analysis, gene ontology, statistical tests, enrichment analysis, non-coding RNA, microRNA.

3.1 Introduction

Ribonucleic acids (RNAs) are important macromolecules which are produced, based on the genomic template, by the cellular process of transcription. The human genome encompasses the templates of approximately 21,000 protein-coding genes (IHGSC, 2004) and numerous functional non-coding RNA genes (Nagano and Fraser, 2011). The former further indicates the assembly of protein using the cellular process of translation. After transcription, human precursor RNAs are further processed and spliced into their mature forms. The mature messenger RNA (mRNA) transcripts include 5' untranslated regions (UTR), 3' UTRs, and the coding region which dictates the translation of proteins.

The latter are exemplified by transfer RNAs (tRNA), ribosomal RNA (rRNA), small nuclear RNA (snRNA), small nucleolar RNA (snoRNA), short interfering RNA (siRNAs), micro RNA (miRNAs), long non-coding RNA and pseudogenes, which perform a wide range of cellular activities. tRNA participates in the protein translation process. Long and short non-coding RNA genes, defined by a heuristic length cut off of 200 bases (Nagano and Fraser, 2011), can regulate other RNAs. miRNAs are a recently discovered class of RNA genes, which regulate many other protein coding genes. The detailed roles of miRNAs within a cell *in vivo* are largely unknown (Cathew and Sontheimer, 2009), a process unveiled recently. For a historical review of the discovery of these RNA genes, please see Eddy (2001).

Unlike the relatively static DNA molecule, whose major function is to pass on information through cell lineages, RNAs can serve as both information carriers and catalytics. RNA is dynamic with its synthesis and degradation regulated by multiple factors and contributes to the dynamics of the cells. Evidence suggests that an ancient primordial RNA world preceded the current system, where DNA is the central inheritance material (Cech, 2011).

RNAs coding for proteins is the central dogma of molecular biology. RNA expression is also regulated by the binding of transcription factors to the promoter region of DNA sequences. The information flow from DNA to RNA, then to protein, gives a simple perception that mRNA and protein expression should have similar time-dependent and tissue-specific patterns. In reality, RNA and protein abundance are not always tightly correlated due to the multiple layers of post-transcriptional regulations and different degradation rates. As a result, the correlation between mRNA and protein expressions is not straightforward.

RNAs have distinct, time dependent and tissue specific patterns of expression. A transcriptomic assay of genome-wide RNA expressions is an essential part of current biomedical research. As Okazaki et al. (2002) put it:

> the transcriptome includes all RNAs synthesized in an organism, including protein coding, non-protein coding, alternatively spliced,

Published by Woodhead Publishing Limited, 2013

alternatively polyadenylated, alternatively initiated, sense, antisense, and RNA-edited transcripts.

A genome-wide exploration offers a global picture, alleviating the bias of previous candidate gene-based studies, which tend to overstate the importance of the candidate genes.

Major directions of transcriptomics studies:

- characterize different states of cells (i.e. development stages), tissues or cell cycle phases by expression patterns;
- explore the molecular mechanisms underlying a phenotype;
- identify biomarkers differently expressed between the diseased state and healthy state;
- distinguish disease stages or subtypes (e.g. cancer stages);
- establish the causative relationship between genetic variants and gene expression patterns to illuminate the etiology of diseases (Schadt et al., 2005).

3.2 Transcriptomic platforms at a glance

Microarrays and deep sequencing technology represent two major technological platforms for exploratory transcriptomic studies. The former is used to quantify the expression pattern of known protein-coding and non-coding RNAs under various conditions, capturing a snapshot of relative abundance of multiple known genes. The latter can be used to detect and quantify both novel and known RNAs.

Microarray technology offers a convenient, high throughput exploration of genome-wide expression patterns of reasonably homogeneous samples. A mixture of multiple cell or tissue types may undermine the specificity and also cause excessive complexity. Microarrays are designed to explore the relative abundance of known transcriptomes in the sample by double-strand hybridization with the oligonucleotide or cDNA probes, fabricated in-situ or attached to solid surfaces for ease of signal detection. Probes usually have fixed lengths, for example, 25 nucleotides (also called "mers") are designed for Affymetrix arrays. However, this is much shorter than an average RNA transcript. Hence, a set of probes (called a probe set or a transcript cluster) are usually designed for a gene, so as to increase specificity (Bolstad et al., 2003). Some probes target the 3'UTR, while others target the exonic regions.

Probes with one nucleotide mismatch have also been designed in some earlier arrays but not in the latest arrays. Therefore, we focus only on perfectly matched probes (ignoring all mismatched probes) in this chapter.

Gene expression microarrays can be broadly classified into two types, the one-channel arrays and the two-channel arrays. A one-channel array is used to assay one sample only. Hence, an exploratory microarray experiment usually requires multiple arrays to process the same numbers of samples under various conditions. A two-channel array is used to assay two paired samples, for example, the cancer and normal tissue of the same patient. Genes in the two samples are attached to two different fluorescent tags during the sample preparation process, resulting in a different fluorescent color on the same array surface. The fluorescent signals indicate the quantity of corresponding RNAs. For a case-control study, a two-color array has an advantage over a one-channel array, in that it offers a doubled sample size in the same number of arrays. It has been shown that both one- and two-channel arrays have equally satisfactory performances in terms of reproducibility, sensitivity and specificity (MAQC, 2006; Patterson et al., 2006).

Prior to a microarray experiment, an RNA extraction step is usually required to prevent other contaminants such as DNA or protein from interfering with the microarray experiment. The same amount of RNA is usually extracted for all samples, for example, 10 ug of total RNA at a concentration of 1 ug/uL. This implies that the microarray experiment only compares the increased or decreased levels of RNA transcripts, assuming the gross transcriptome amounts are the same under all conditions. Obviously, this assumption only partially approximates to the real situation. Hence, the global measurement and comparison of ups and downs of gene expression come at a cost of relative measurements. The measurement of absolute quantities can be achieved by other technologies, such as qRT-PCR or nano-strings (Geiss et al., 2008). This equal-quantity assumption affects the analysis procedure, such as the normalization step, which will be detailed later.

Recently, RNA deep sequencing has been utilized to play the role of microarray assays on the quantification of RNA abundance or concentration, which is estimated by the depth of transcripts (Gamsiz et al., 2012). This is still a relative measurement. RNA deep sequencing offers a few advantages over conventional microarrays, particularly in the detection of unknown transcripts or rare alternative forms, and sensitivity on low-abundance transcripts. Hence, this technology is useful for characterizing alternative splicing patterns in different cell states.

Published by Woodhead Publishing Limited, 2013

Polymerase chain reaction (PCR) technology enables RNA transcripts to be amplified multi-fold to the desired level of detection. Based on this, the quantitative real-time PCR (qRT-PCR) is a common technology for measuring RNA abundance of individual genes. This technology measures RNA levels during cycles of PCR. A higher original amount of RNA in the original sample will mean the PCR product is saturated (plateau) early. Thus, the number of cycles before saturation can reflect the original RNA amount. It should be noted that RT-PCR offers only a "relative" measurement. A comparison with standards or calibration curves is usually required for their absolute quantification.

3.3 Platform level analysis for transcriptomics

A transcriptomics dataset, produced by either next-generation sequencing (NGS) or microarray platforms, usually comprises the quantitative measurement of thousands of genes, a desirable feature for exploration of mechanisms. The power of such a dataset can only be revealed when an adequate data analysis is in place. *Post hoc* analysis comprises three major steps in the sequence: platform level processing, contrast level data filtering or extraction, and module level characterization of behavior or synergistic effect of groups of genes. As will be seen in the following case studies, a variety of methods are available, so that an exploratory analysis remains an art rather than a standard operating procedure.

The direct output of a microarray reader is a scanned digital image. The platform level analysis, also known as the pre-processing step, aims to prepare the data so that it is presented in good-quality, convenient, spreadsheet-like formats for subsequent analysis. This is usually well supported by vendor software accompanying the technological platform. Such software is generally reliable and straightforward to use. Here we will describe only the principles of platform-level analysis.

3.3.1 Background subtraction and correction

These are important steps to adjust the intensity values based on local, nearby intensity distribution of the image. The goal is to remove bias originating from non-specific bindings of RNAs, and the uneven background fluorescence, both of which can be estimated using local intensity profiles. The intensities in the proximity of the detection probe may be characterized by two models, one for the background and one for the detected signal of probe hybridization. The probe intensities are then corrected by subtracting the estimated local background levels. This two-model method can offer a binary reading of gene expression, that is, the presence and absence of intensity, if such binary information is required instead of quantitative measurements. The probe intensities, which fit into the background model, are declared as "transcript absent". Otherwise, they are "transcript present".

3.3.2 Summarization

The second step aims to summarize or average the intensities to give an abundance estimate per higher-level unit such as gene, probe set or transcript cluster. Whether a summarization or an averaging is adequate depends on the relationship of probes. If two probes are designed to target two alternatively spliced forms of a gene, then a summarization may be more adequate to represent the total amount of the genes in the two forms. If different number of probes are used to capture a single gene transcript, for example, five probes for gene A and ten probes for gene B, then an averaging may be more suitable, which does not overestimate the abundance of B due to more probes being used.

3.3.3 Normalization

The key purpose of normalization is to correct systematic errors caused by non-biological error sources, and ensure a fair comparison of biological effects. This step is related to the array types because of different error sources. For one-channel arrays, the critical goal is to remove variation across chips. Several scale normalization methods, such as median polishing or quantile normalization methods, are commonly used (Irizarry et al., 2003b). The intensities of chips are adjusted into similar distributions so as to ensure a fair comparison. This is particularly useful when technological replicates are employed, which by definition do not bear distinct biological effects. Scale normalization is exemplified by the popular Robust Multi-array Average (RMA) algorithm, a scale normalization based method which has also integrated a summarization step (Irizarry et al., 2003a). It is worth noting that the output value of normalization is in the range of the logged 2 value of original intensities, under the assumption that the logged

Published by Woodhead Publishing Limited, 2013

value better represents the relative abundance of genes. This algorithm has been implemented into several free software packages such as the LIMMA package in R. RMA has also been implemented in a window-based software called RMAExpress (Bolstad et al., 2003) (Box 3.1).

Box 3.1: Resources and tools for transcriptomics data analysis

A. High quality transcriptomics data resources

- GEO

GEO stands for Gene Expression Omnibus, a public repository of microarray data hosted by the NCBI site. Currently, it is common practice to publish the microarray raw data to accompany a research publication for public access. GEO has become a *de facto* choice of repository due to its publicity. It is also a valuable source of high-quality raw data. Users often use keywords to identify the microarray study they are interested in. Or alternatively, if they have obtained the GEO ID from literature, they can use the ID to query the relevant dataset.

The website also offers several data analysis tools. Amongst them, GEO2R can perform two-group contrast level analysis directly using the deposited data. Box plots of the intensity distribution can be presented visually. GEO2R is based on the LIMMA (based on R) analysis package. It can generate the scripts of R commands for user reference. An efficient use of GEO2R can drastically save time and effort for two-group comparison. With this software, the chore of downloading the raw data can be avoided for exploratory purposes, unless a meta-analysis of multiple datasets, or a special type of calculation such as ANOVA, is required.

- ArrayExpress

ArrayExpress is a public repository of microarray data hosted by the European Bioinformatics Institute (EBI) site. This is also open for scientists to submit array data, and conduct analysis. A meta-analysis was also conducted by a team of curators to produce the Gene Expression Atlas, offering a graphical summary of gene level ups and downs in different organs and conditions.

B. Contrast-level analysis

- The LIMMA analysis package

The LIMMA analysis package is dedicated to microarray data analysis. It is constructed on top of the statistical package R, and is distributed along with the Bioconductor package. It is the basis of the GEO2R tool provided by the NCBI site. The LIMMA package can be used to load data files of a variety of vendors such as Affymetrix and Agilent. It has two companion graphical user interfaces (for one- and two-channel microarrays, respectively) to facilitate data manipulation. It can also perform normalization by RMA/GCRMA methods, as well as contrast level analysis of two groups of samples.

To use LIMMA in local computers, the user needs to install in sequence the R statistical package (from the R project website) and Bioconductor (from the bioconductor website). Graphical user interfaces are then installed.

- The Significance Analysis of Microarray (SAM) method and package

SAM is a non-parametric, permutation-based method proposed specially for microarray data analysis (Tusher et al., 2001). It calculates the empirical False-Discovery Rate (FDR) by the random permutation of class labels. The permutation generates a null distribution, because the randomness is assumed to remove all biological effects. Therefore, it provides a means to control the false positives under various thresholds when multiple genes are assayed simultaneously in an array. The SAM package can handle both paired and non-paired data. It is run on top of the R statistical package, and has an excel interface using an excel plug-in.

- Past

Past is a free statistics software, which can be used for a variety of univariate and multivariate analyses. It has a spreadsheet-like

Published by Woodhead Publishing Limited, 2013

interface for data input. It can also be used to generate plots, such as a PCA plot.

C. Module-level and Integrative analysis

- Cluster and TreeView

Cluster and TreeView are commonly used for clustering-based module-level analysis. The current version Cluster 3.0 can perform various clustering methods such as hierarchical and k-means clustering. The clustering can be done by gene, by samples, or both. It also offers several ways to quantify the similarity of two gene or sample vectors, such as Pearson's correlation, Spearman's rank correlation and Euclidean distance. The result can be visualized using TreeView, which is an open source software hosted by Sourceforge.net. Users can select different colors for data visualization in TreeView. Both Cluster and TreeView can handle missing data, a useful feature for practical data analysis.

- PANTHER

PANTHER is a web service particularly useful for module level annotation. The user can provide one or two lists of genes, and the software can calculate enrichment based on gene ontology classifications or protein classifications provided by the site. The functional classification result can be presented graphically in several ways; all of them are excellent.

- GSEA

Gene Set Enrichment Analysis (GSEA) was developed by the Broad Institute in MIT. This software does not rely on any particular cut off of statistical significance in contrast level analysis. It employs the rank of significance instead and determines the enrichment of top listed genes to a biologically meaningful set of genes.

- EnrichNet

EnrichNet is a server of network-based gene set enrichment analysis hosted at the University of Nottingham, UK. The server contains several networks including Biocarta, KEGG pathway,

Reactome and Gene ontology, which are readily available for users to search. A novel algorithm with a network-based score was implemented for searching (Glaab et al., 2012).

- DAVID website

DAVID is provided by the National Institute of Health. It offers an integrated annotation combining gene ontology, pathways and protein annotations.

- GOEAST

GOEAST is an enrichment analysis tool based on gene ontology. Enrichment is based on the hypergeometric test. It offers a tree-like presentation.

- GeneXPress

GeneXPress is a software developed for heat map visualization and annotation (mainly for yeast). It is written in the JAVA language.

- Function Express at Washington University

This is basically an annotation suite. An application via email is required for using this software.

For two-channel arrays, such as Agilent arrays, a major source of error is the dye-associated intensity bias. This is caused by either different binding affinity between dyes and samples, or unequal dye luminance. Dye-associated bias may result in detection of genes not associated with the study purpose. As two-channel signals are often analyzed in pairs (e.g. analyzing red and yellow signals together by their log ratios, or by paired statistical tests), a per-array normalization such as Lowess may be adequate (Kerr 2007). Lowess stands for locally weighted least squares regression. The basic assumption is that two signals (by the two dyes) should have a balanced up and down in every signal range (low, middle or high), where the signal of a gene is estimated by the geometric mean of the two signals in a pair. This method estimates the degree of imbalance (upward to downward) within a signal window (a range of signal), then tries to offset the imbalance in the window to produce a normalized signal. Apart from Lowess, we can also use stronger normalization methods across all arrays in a study, such as RMA offered by the J-Express software (Dysvik et al., 2001), to handle two-channel array data.

Published by Woodhead Publishing Limited, 2013

It is worth noting that excessive pre-processing may introduce unwanted artifacts into the analysis. For example, it has been shown empirically that the background subtraction step may introduce problems to two-channel platforms by exacerbating the noise variation of low intensity genes (Qin et al., 2004). The adequate degree of correction and normalization actually depends on the data quality and the purpose of the study. Data with poorer quality often needs more pre-processing.

3.4 Contrast level analysis and global visualization

The major purpose of contrast level analysis is to delineate the pattern and quantity of gene level difference (i.e. the contrast of gene levels) of cells under different conditions, for example, under an intrinsic change of cell states, or extrinsic stimulations such as by drugs or stress (i.e. the contrast of cellular conditions). Contrast level analysis needs to be in line with the study design. Global visualization is critical for illustrating expression profiles before or after contrast level analysis, offering tremendous insights into the properties of data.

3.4.1 The data matrix and the heat map

It is imperative to first organize the complete set of transcriptomic data, often obtained on different days or from different laboratories, into a manageable format for data analysis and visual presentation. A gene-by-sample matrix is often useful, with row vectors of RNA levels associated to a gene, a probe set or an individual probe. Here we will use a gene per row to simplify the discussion. The column vectors are expression levels of individual samples (Eisen et al., 1988).

Mathematically, we can denote the number of genes as m, the number of samples as n, then the gene-by-sample matrix as an m-by-n matrix. Denote the index of samples as j such that $1 <= j <= n$, and denote the index of genes as i such that $1 <= i <= m$. A sample is represented as a column vector of quantitative measurements:

$$Xj = [x1j; x2j; x3j; x4j; \ldots\ldots xmj].$$

Similarly, a gene is then represented as a row vector of quantitative measurements:

$$Yi = [yi1; yi2; yi3; yi4; \ldots\ldots yin].$$

By aggregating batches of samples into a large matrix, we obtain the matrix D:

$$D = \{Xj| j = 1:n\} = \{Yi|i = 1:m\}.$$

The data matrix can then be visualized using a spectrum of gray shades or colors, representing the relative abundance of genes. This presentation is called a heat map (Figure 3.1), an extremely useful tool for data interpretation and exploration. A gray scale or color bar is often provided to accompany the heat map for the interpretation of relative gene levels. The gray scale or colors can be adjusted per gene across all samples to suit different purposes of presentation. For example, it can be adjusted so as to reflect the elevation or suppression of gene levels compared with a control condition (i.e. the baseline of medical treatment). This can be done by subtracting control gene levels across all samples, resulting in the mean value of the control group as 0. One alternative method is to adjust the levels so that the mean of each gene becomes 0 and the standard deviation becomes 1 (i.e. with unit standard deviation). This can be done by subtracting the average gene level of each row across all samples. Such an adjustment does not affect the correlation coefficients between genes.

Despite the similar appearance, RNA data matrices are distinct from DNA data matrices in light of the following aspects:

- Data associated to a subject are now represented by a column vector, rather than a row vector as in the DNA matrix.
- Each cell now contains a quantitative, floating point value representing the gene expression levels or their ratios, rather than the categorical data of genotypes.
- The gene-by-sample matrix is commonly visualized using the heat maps, where the quantitative values are shown by a spectrum of colors.
- The genes are not required to be sorted according to the chromosomal positions; instead they are usually sorted by the similarity of gene expression patterns, so as to visualize the hidden modular activity in the heat map.

The sorting is done by methods such as hierarchical clustering or self-organizing maps (see Section 3.5.1 below on "similarity clustering"). In such cases, the hierarchical structure of genes may also be shown on the side of the heat map (Figure 3.1).

Published by Woodhead Publishing Limited, 2013

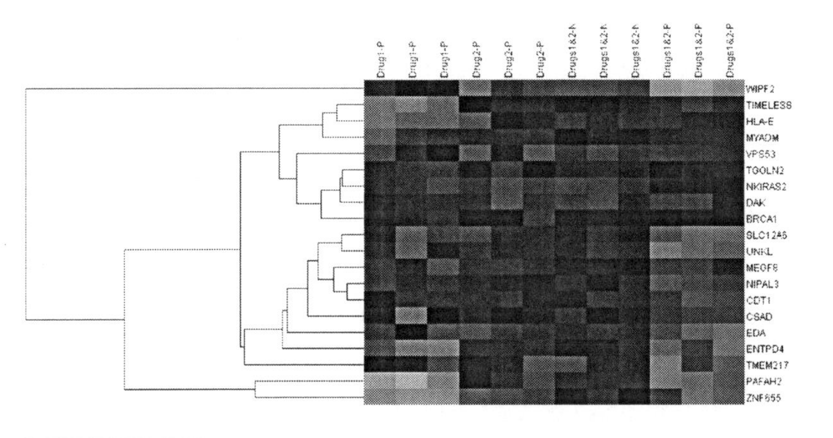

Figure 3.1 Figure 3.1 A heatmap visualization of a gene-by-subject expression datamatrix. Each row represents a gene and each column a subject. The characteristics of subjects are usually indicated above the columns. In this example, subjects were treated (P) or not treated (N) by Drugs 1 or 2. Three replicates were used for each treatment. The gene name is usually labeled to each row, on the right side of the matrix. A hierarchical tree is shown on the left side of the matrix to show the correlation of gene expression levels. Genes with similar expression profiles (in terms of high correlations) are presented closer to each other. Gene expression values are shown as gray scales. This image was generated by Cluster 3.0 and TreeView (Box 3.1)

3.4.2 Contrast detection and the volcano plot

Contrast detection aims to find genes (probe sets) amongst all the assayed genes, which exhibit strong gene level contrast across distinct experimental or clinical conditions. In the presence of inevitable biological variation and noise, contrast detection needs to consider two issues:

1. The contrast is not likely to be false positive, a bottom line of an exploratory result.
2. The contrast should be large enough to bear biological functions.

Parametric statistical tests are commonly used to estimate the significance level, called P-values, of contrasting genes of two distinct experimental conditions. The design of the experiment can determine whether the samples are paired or non-paired. The paired samples are often two samples of the same subject at different time points (e.g. before or after medication) or from different tissues (i.e. tumor sample and adjacent healthy tissue). These samples can be compared by parametric, paired t-tests, which alleviate the cross subject heterogeneity and focus on the experimental contrast. However, the non-paired t-tests are used to assess the group average of experimental conditions.

The equation for non-paired two sample t-test (with unequal variance) is shown here to illustrate the principles of contrast detection (Rosner, 2006):

$$\mathrm{t(g,c)} = \frac{x1(g) - x2(g)}{\sqrt{\frac{s1^2}{n1} + \frac{s2^2}{n2}}}$$

where g represents gene expression levels and c = {1,2} represents two distinct classes. n1 and n2 are the number of samples in the two groups, x1(g) and x2(g) represent the means of the two groups, and s1(g) and s2(g) represent the standard deviations of the measured data of the two groups respectively. The P-values, which represent the probability of false positive when there is no biological effect, can be derived based on t(g,c) and the degrees of freedom d:

$$d = \frac{(\frac{s1^2}{n1} + \frac{s2^2}{n2})}{(s1^2 / n1)^2 / (n1 - 1) + (s2^2 / n2)^2 / n2 - 1}$$

A similar index of contrast is called the signal-to-noise ratio (SNR), also frequently used for microarray data analysis. The contrast of gene expressions is termed "signals", while the standard deviations of the gene expression levels are termed "noise". The equation is given as

$$P(g,c) = [\mathrm{x1}(g) - \mathrm{x2}(g)]/[\mathrm{s1}(g) + \mathrm{s2}(g)]$$

Comparing the equations of SNR and t-tests, difference of mean expressions of the two groups are both presented in the numerator, and the standard deviations are both in the denominator. The idea is to select genes with large differences and small variations. SNR-based methods usually rely on the permutation test so as to generate an empirical null distribution. The significance level is then calculated based on the null distribution.

Published by Woodhead Publishing Limited, 2013

Often microarrays are used to explore expression patterns in more than two experimental or clinical conditions. In such cases, more elaborate methods than t-test or SNR are required. A variety of analysis strategy can be used, depending on the study purpose. First, the analysis of variance (ANOVA) test can test whether a gene has different expression profiles across various conditions. Second, several conditions may be regarded as the same and the samples aggregated as one group. For example, patients of Type 1 diabetes and rheumatoid arthritis may be pooled together as one group to be compared with healthy controls, based on the rationale that they both belong to autoimmune diseases (WTCCC, 2007). Third, if the clinical conditions are marked by levels of severity, then we may employ ordinal regression. For example, we may want to capture the molecular signature of cancer stages, or the five liver fibrosis stages F0 to F5 reflecting varying severity of the disease. Ordinal regression is an extension of logistic regression, which will be discussed in Chapter 4. Finally, a combination of the above strategies may be employed to form multiple constraints; for example, if the samples are of A, B and C conditions. We may select genes with a significant difference between A and B, as well as A and C. The multiple constraints aim to harvest genes to simultaneously fulfill all the constraints.

To adequately evaluate the contrasts, it is suggested that each group needs to have more than three biological samples. One major reason for the lower limit of sample size is to adequately quantify the variance (or standard deviation) of the expression levels.

The aforementioned parametric tests, such as t-tests and ANOVA, both assume the data are sampled from normal (i.e. Gaussian) distributions. In practice, the expression levels of genes often have skewed distributions across samples, many samples exhibit a relatively low value of expression levels toward zero, and a minority of samples demonstrate a very high value. The data distribution has a long tail toward the high value, clearly not a normal distribution. When the assumption of data normality is in question, the non-parametric methods can then be used (Mann and Whitney, 1947). The Mann–Whitney U tests are non-parametric alternatives to non-paired t-tests. The Wilcoxon signed-rank test is an alternative to paired t-tests. We can examine the data using methods such as the Lillifore test to check whether the data normality assumption is valid (Rosner, 2006).

Genes can then be ranked by their significance levels, which are the probability of false positive. In addition to the evaluation of significance levels, we also need to select the genes with larger expression level differences or folds across conditions, because they are more likely to bear salient biological functions. Scientists have selected genes based on fold changes, differences of means, SNR, P-values or their combinations. The combination of criteria can be illustrated by a volcano plot, which is a useful scatter plot for presenting the results of various contrast level analysis. It is particularly useful to show the combined gene selection criteria when multiple experimental conditions are assessed (Figure 3.2). The vertical axis usually indicates the negative log P-value. Genes with higher significance levels (i.e. smaller P-values) are shown in the top area of the plot. The horizontal axis can be used to represent the second criteria, such as the logged fold change, which is a common gauge of association used by many scientists. It can also represent the mean difference of the two conditions and the slope of a linear increase or decrease along the time axis or dose scale. An important consideration about the horizontal axis is to have a symmetrical scale for ups and downs. That is why the fold change value, the ratio of two expression levels, needs to be logged. The volcano plot is also useful for the presentation of genomic (DNA) variant association studies, where the vertical axis is still the logged P-value and the horizontal axis is the logged odds ratios.

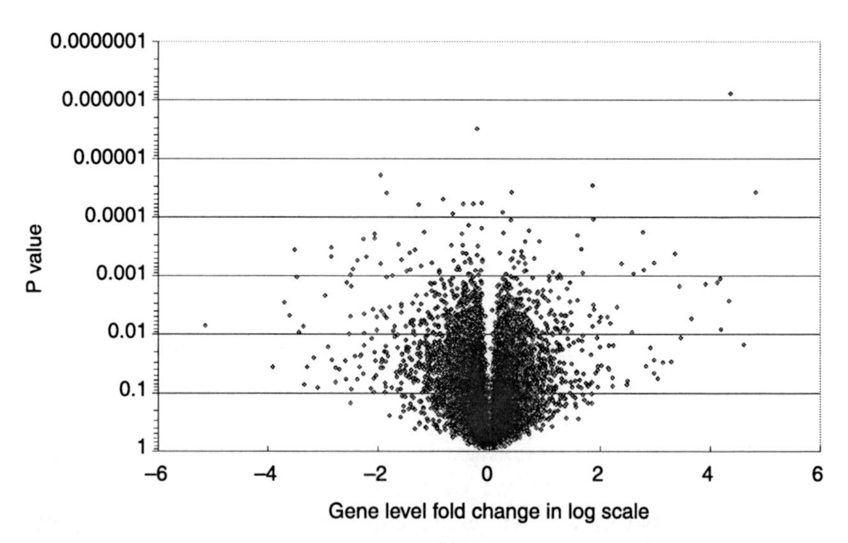

Figure 3.2 A typical volcano plot of a transcriptomics study. The *x*-axis shows the logged gene level fold change of case and control groups. The *y*-axis shows the negative log 10P value (This plot was generated by Excel.)

Published by Woodhead Publishing Limited, 2013

The different selection criteria will inevitably render different results. It has been an issue of debate whether the difference of means or fold change should be used for selection. Some prefer absolute differences (Tusher et al., 2001). MAQC prefer fold change rather than P-values, for reasons of consistency between studies.

Contrast level analysis can be calculated by spreadsheets and a variety of open sources, freeware (e.g. LIMMA and SAM (Box 3.1)) or commercial software available for microarray analysis, as well as statistical packages. Non-parametric tests can also be performed by a variety of statistical packages, such as SPSS.

3.4.3 Principal component analysis and PCA plots

Principal component analysis (PCA) and independent component analysis (ICA) are useful mathematical tools, which can generate plots to present the distribution of samples. In a transcriptomic data matrix, samples are characterized by levels of multiple assayed genes. These genes span a hyperspace of multiple orthogonal dimensions, one gene in each. Samples are pinpointed in the hyperspace based on the gene levels. For ease of comprehension, presentation of samples often requires dimension reduction methods. The PCA detects a 2-D or 3-D cross-section of the hyperspace, on to which the samples are projected (Figure 3.3).

PCA and ICA are mathematical tools for the reduction of multivariate, high dimensional data. PCA and ICA are the top-down construction of representative vectors, and reduce high dimensional data into a presentation of fewer representative dimensions. Based on the data distribution in high dimensional space, PCA analysis extracts a series of principal components (linear transformed coordinates), where data in the first principal component has the largest variant. The second principal component is perpendicular (orthogonal) to the first principal component and has the second largest variant. The underlying assumption is that the coordinates with the large variants most saliently demonstrate the contrast between sample points, while the coordinates with smaller variants may be a source of noise, which should be ignored or suppressed. In the meantime, the correlation between two dimensions represents redundant information, which will not be presented. That is why this algorithm requires the following coordinates to be perpendicular (orthogonal) to previous coordinates. A PCA analysis can reduce the

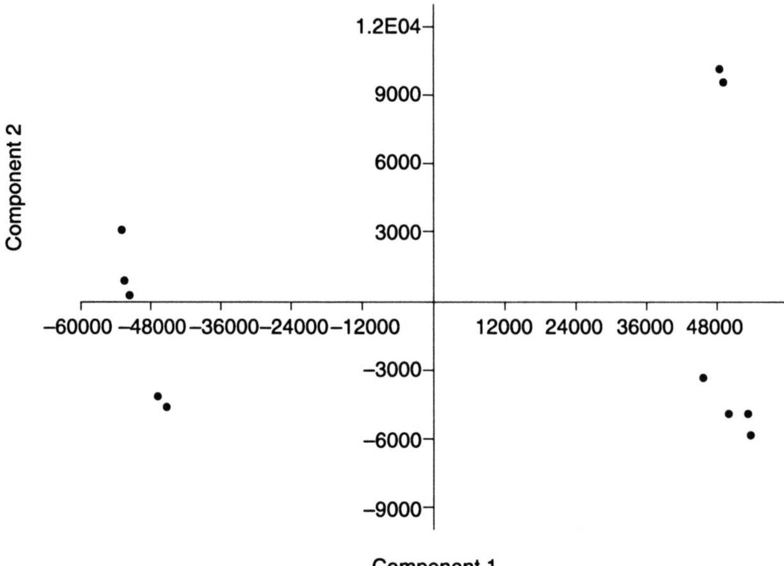

Figure 3.3 A 2-D plot by principal component analysis. Eleven samples are represented by the dots. The first and fourth quadrants represent samples of two different phenotypes. The dots in the second and third quadrants represent samples of one other phenotype. (This plot was generated by PAST software (Box 3.1).)

dataset of m dimension into n dimension, $n <= m$, by selecting the first n principal components.

However, ICA aims to decouple signals with distinct sources. This is also known as the blind source separation for solving the well-known cocktail party problem: detect each single conversation from a loud, noisy cocktail room. As opposed to PCA, the ICA does not utilize the orthogonal assumption that each representative component needs to be in the perpendicular. PCA and ICA can be performed in sequence. PCA is used first to extract the n principal components, then ICA is put in place to adjust the components so as to relax the orthogonal assumptions.

In transcriptomic data, samples are column vectors of m-dimensions, where m is the number of gene measurements. The samples are scattered in an m-dimensional vector space, where each gene is a dimension (or a coordinate). Each sample contributes a sample point (a dot) to the space. The sample vector is a vector linking the origin of the coordinate system toward the sample point.

Published by Woodhead Publishing Limited, 2013

3.4.4 Other global visualization methods

We have already seen the usefulness of the heat map for global visualization of gene level ups and downs under various experimental conditions (Figure 3.1). One additional useful class of plots is called 2-D scatter plots, which can be adapted for different purposes. Dots in a scatter plot represent a data point of a gene (or probe set). First, it can be used to show the dye intensity bias for two channel arrays, which is useful for platform level analysis. Second is the ratio intensity plot, also known as the M vs. A plot, where M is the fold change of two conditions. The fold change can be defined in one of several ways when more than two samples were measured, for example, the average of the paired fold, or fold change of the unpaired case and control samples averaged separately. The value is shown on the vertical axis. However, A represents the average gene expression level (e.g. the geometric means) shown on the horizontal axis.

3.5 Module level analysis

3.5.1 Similarity clustering – forming modules

Similarity clustering is an important step in the exploration and organization of expression profiles into modules. It can be used to group genes with similar expression profiles across samples, or samples with similar expression signatures across genes. Genes with similar expression profile are assumed to be co-expressed together by the same set of transcriptional factors, or for serving closely related cellular functions. The similarity of gene expressions makes it difficult to delineate which gene "causes" the expression of other genes. However, from another perspective, the similarity of gene expressions indicates that they can be analyzed together as a module to simplify the analysis. However, samples with similar expression profiles may represent the same clinical phenotypes, the same degree of external stimulation or drug treatment, or the same subtype of a heterogeneous disease. Similarities and differences are two sides of the same coin. For example, if the difference between two samples is defined as the Euclidean distance of their gene level vectors, then a smaller distance indicates a higher similarity.

This method has been used alone or in conjunction with contrast level analysis for drawing conclusions (Golub et al., 1999). If done alone, then the entire gene-by-sample matrix is clustered into modules, as long as the computational resource can process that much data. Modules of samples can represent subtypes of phenotypes such as cancer types, which are defined by all the gene levels of the transcriptomic data. If it is done before contrast level analysis, then the modules produced are further evaluated by their group contrasts, for example, the average of all genes across experimental conditions. If it is done after contrast level analysis, then the selected genes with significant differences are further grouped into modules, and each may represent an activated or suppressed cellular function. These methods can be used flexibly, depending on the purpose of the study. A heat map presentation usually is preceded by a similarity clustering for ease of comprehension and interpretation.

The clustering of samples by their similarity of expression vectors has been shown to be useful for disease sub-typing (Sarwal, 2003; Bhattacharjee et al., 2001). An unsupervised method was conducted, where the grouping was based purely on the gene expression signatures without the disease subtype of the samples. The disease subtype or clinical feature of each sample was known (by pathology tests), but temporarily undisclosed, so the analysis was done as if the subtype was unknown. The grouping result by unsupervised clustering was shown to be in good agreement with the known disease subtypes. Hence, it is certain that the molecular signature will be able to represent the macroscopic phenotype.

Clustering methods can be categorized as single layer or hierarchical methods, the latter being further divided into bottom-up or top-down approaches. The agglomerative hierarchical clustering method is a bottom-up process, where genes (or subjects) are progressively merged, based on the pairwise similarity between each pair of genes (subjects). Here the similarity of two vectors is often gauged by their pairwise Euclidean distance, Pearson's or Spearman's correlation. Hence, all genes (subjects) have equal opportunity to be merged with other genes. The Unweighted Pair Group Method with Arithmetic (UPGMA) mean method is an example of the bottom-up hierarchical clustering method. A tree-like hierarchical structure is progressively constructed, where the root of the tree represents a single large cluster containing all genes (subjects). Based on the tree, this cluster can be partitioned into several smaller, relatively homogeneous clusters, in the sense that the similarity between the cluster members is relatively lower. As we trace from the root further down, more cluster will be generated, and each cluster will be more homogeneous. Thus the number of clusters is adjustable by the investigator for interpretation of the data. Each cluster can be seen as a module, as the genes (subjects) within a cluster share a similar expression profile, implying their similar cellular role (for a gene module) or disease

Published by Woodhead Publishing Limited, 2013

subtype (for a sample module). The Cluster 3.0 software (Eisen et al., 1998; de Hoon et al., 2004) is a widely used, public domain software for hierarchical clustering (Box 3.1). Its results can be conveniently loaded into the visualization software TreeView (Saldanha, 2004) (Box 3.1).

The single-layer clustering methods are exemplified by the k-means clustering and self-organizing map (Chaussabel et al., 2008; Golub et al., 1999). K-means clustering is an iterative process where the members associated to the clusters, and the cluster centroids, are progressively updated until a stable state is reached. This requires the vector to be standardized to a zero mean and unit variance. However, this method requires the number of clusters to be determined before analysis. Global clustering methods are useful for time-course analysis where genes co-expressed in time result in similar time-course profiles, which are then grouped together.

It is also commonplace to have arrays assayed at different time points. This type of experiment often requires clustering of genes of a similar time course profile. In such cases, various clustering techniques can be used. For example, k-means clustering has been employed in time course data of neutrophil activation experiments (Kotz et al., 2010).

We can add some twists based on the same concept of grouping of genes into modules to extend the range of applications. Instead of grouping directly, based on the similarity of expression patterns, we can also base on the frequency of oscillatory gene expression, which may have different phases (Kim et al., 2008). Genes with the same frequency but with different time lags may actually serve the same periodic cellular role.

3.5.2 Functional annotations of activated biological modules

Contrast level analysis renders a focused set of genes or probe sets of interest, which show stronger associations with the contrast of experimental conditions. Once the subset of genes is identified, the next step is to explore the functional significances of the genes individually. The first step is to extract preliminary annotations of each gene in the subset via microarray vendor websites (i.e. Affymetrix NetAffx website), or other online databases (i.e. DAVID website (Box 3.1)). Annotations obtained in this way usually include a fairly general description of the gene, such as its gene ID, symbol, sequence ID, location in the genome, and a few lines of summary of functions. This information is often too preliminary to conclude the experiment. One solution is to query a multiple database so as to produce a tailor-made annotation for a particular study.

Preliminary annotation can be quickly obtained; however, the real challenge lies in the module-level annotation to partition, filter or organize the short-list of genes into functional modules so as to account for the total experiment. This is probably the most challenging and critical step before forming a theory and drawing a conclusion.

Enrichment analysis is a frequently-used approach for functional annotations. This is done by matching the focused set of genes to all known pathway modules (e.g. the extrinsic apoptosis pathway, the renin-angiotensin pathway, etc.) or gene ontology modules with respect to molecular functions, cellular components or biological processes. The analysis detects modules with matches of genes significantly higher or lower than expected values, showing that the focused set of genes is enriched or deprived in the module of interest. The significance is often assessed by the hypergeometric test. Assuming the total number of genes in the human genome is N and those in that module of interest is m, and assuming the number of genes in the focused set is n and the number of matches is k, then the probability of k in the hypergeometric distribution is:

$$P(k) = c(n,k)c(N-n,m-k)/c(N,m)$$

An up- and down-regulated subset of genes can be submitted for enrichment analysis together or separately. It is often easier to do them separately.

The underlying assumption behind the enrichment analysis-based annotation is that functions of active modules (either boosted or suppressed) may manifest by the distribution of constituent genes. Hence, a module enriched or deprived with the genes in a pathway suggests that the pathway may play a salient role in the topic under investigation. Gene vs. module matching offers a type of domain knowledge filter to sift important genes segregated into modules from a background of noise. However, this assumption is not always true. In practice, the focused set of genes often scatter in different directions. Those genes not enriched in a module are left unexplained, despite the fact that they may actually play key roles. They could either work alone, or synergistically, in a way we still do not understand and thus are not enriched in any particular pathway (yet). This is one major drawback of the use of enrichment analysis at the present time.

Therefore, the success of enrichment analysis highly depends on the accuracy and completeness of the genes in the pathway database. As the pathway information is still incomplete and new knowledge is

Published by Woodhead Publishing Limited, 2013

accumulating fast, using software with the most recent pathway information will hopefully improve the quality of functional annotation. In addition, pathway usually contain a smaller number of genes than a biological process, and the former may offer high specificity but lack sensitivity. This suggests that we should conduct functional annotations using both types of module so as to better understand the result.

A variety of online software tools for enrichment analysis are currently available. The PANTHER (Protein ANalysis THrough Evolutionary Relationships) Classification System is a valuable functional annotation website, which offers a global presentation for the comparison of two focused sets of genes, an ideal feature for presenting both the up- and down-regulated genes (Thomas et al., 2003). The GOEAST (Gene Ontology Enrichment Analysis Software Toolkit) website provides a convenient tree-structure presentation (Zheng and Wang, 2010) (Box 3.1). The DAVID site by NIH is very useful as it provides an integrated analysis of gene ontology categories and pathways. In addition, many gene ontology packages are available as plug-ins to the Cytoscape software, a graph-centered software platform (Box 5.1). These packages are excellent for building network presentations of gene ontology terms. GoMiner and GenMapp are useful software for annotation analysis.

Metacore is a commercial software package for module level functional annotations in the pathway or gene ontology analysis. One special feature of the software is the ability to build a set of genes into a network. Genes are linked if there is a literature reference connecting the two genes. Needless to say, this relies on the literature mining done by the company.

3.5.3 Gene expression matching

One other approach of module level analysis is to match the focused gene of one experiment directly to another experiment. A good match indicates that the underlying mechanisms behind the two experiments are similar. This is actually engineering style thinking to regard a complex, isolated system (in this case, an experimental perturbation of genome-wide RNA levels) as a "black box", and then try to characterize the system by inputs and outputs. The same outputs indicate the same inputs. This idea also applies to "opposite" gene expression alteration pattern (a reverse of up and down), which suggests an antagonizing effect. If a drug produces an opposite expression pattern of a disease, then the drug may be a good therapeutic agent for the disease. As a result, the gene expression pattern becomes a common intermediate across different experiments.

To put this concept into practice, a reference dataset needs to be established first. Lamb et al. (2006) conducted a series of *in vitro* microarray experiments on cell lines stimulated by various currently used drugs, providing the reference dataset for gene expression matching. Microarray experiments were conducted before and after drug stimulation. The reference database is called the Connectivity Map (Box 3.1).

A focused set of genes can be submitted to the Connectivity Map server. The direct and opposite matches indicate the clues of mechanisms behind the gene expression alterations of the focused set of genes. The server employs the Kolmogorov–Smirnov statistic-like scores to handle the matches. First, up- and down-regulated genes are matched separately, then combined together for ranking of similarities. The Kolmogorov–Smirnov test is a non-parametric method to test the similarity of distribution functions. The server also uses permutation methods to calculate significance levels of matches (i.e. P-values).

3.5.4 Gene Set Enrichment Analysis

The Gene Set Enrichment Analysis (GSEA) algorithm and software, offered by the Broad Institute, represents another class of module level analysis (Subramanian et al., 2005; Mootha et al., 2003). The goal is to analyze the focused set of genes to see whether they are the leading perturbed genes amongst all the genes. P-values are derived from an empirical distribution of permutations of the class labels.

3.6 Systems level analysis for causal inference

Systems level analysis refers to the joint analysis of multiple systems, such as the genomics and transcriptomics. Coherent evidence derived from different systems can enhance the evidence level and the confidence. A joint analysis involves both integrators and filters, an analogy drawn from common signal processing devices seen in digital circuits (Ideker et al., 2011). Contrast and module level analysis can be seen as filters to remove biologically irrelevant noise from true signals. Now we will explore the integration of multiple datasets.

Each individual discovery of association between germline, somatic DNA variants or RNA expressions and a diseased state is a milestone.

Published by Woodhead Publishing Limited, 2013

Often the association is based on the grounds of statistical associations rather than mechanistic explanations. Not every genetic variant associated to disease reflects the cause of the disease, because many adjacent DNA variants are correlated with each other, known as the linkage disequilibrium phenomenon (Chapter 2). In such cases it is difficult to distinguish the real causative SNP from the adjacent, statistically associated but not causative SNPs. Similarly, not every gene expression associated with the disease reflects the cause of the disease, because the expressions of multiple genes are usually correlated to each other due to their intricate regulations, which might not be directly responsible for the trait.

Once a germline or somatic variant is statistically confirmed to be associated to a disease, the underlying mechanical cause may be hypothesized, depending on the location of the variant in relation to a gene:

1. If the variant is in the coding region of the exon, and the different alleles code for different amino acids (called a non-synonymous variant), then the amino acid variant may be responsible for altered protein structure or function efficiency, which may be the cause of the disease. This hypothesis can be further validated by protein assays.
2. If the variant is in the coding region of the exon, but different alleles do not code for different amino acids (called a synonymous variant), then it may be hypothesized that this variant will change translation efficiency.
3. If the variant is in the exonic UTR, such as 3'UTRs and 5'UTRs, then it may be postulated to change RNA regulations.
4. If the variant is in the intronic region, then it may be postulated to cause RNA splicing patterns. This will further cause protein level change. This hypothesis can be substantiated by checking the relationship between variant and mRNA alternative splicing forms.
5. If the variant is around the promoter and transcription element region, then this variant may be implicated on transcriptional efficiency, therefore affecting the RNA (and then protein) expression pattern.

Each hypothesis can indicate subsequent designs of validation studies to illustrate the mechanical cause (Peer and Hacohen, 2011). For scenarios 3, 4 and 5, a joint analysis of genomics and transcriptomics is justified. Gene expression has been shown to correlate with GWAS studies under certain conditions (Gorlov, 2010). A causal, mechanistic relationship between genes and phenotypes under scenarios 3 to 5 includes three criteria:

A. Genomic variants associated to RNA expression;

B. Genomic variants associated to phenotype;

C. RNA expression associated to phenotype.

With the transcriptomics platforms such as exon arrays or expression arrays in place, an integrated study design and concurrent analysis is motivated (Schadt et al., 2005). Criteria B and C are first examined individually by genome-wide contrast level analysis. DNA and RNA are preferably measured in the same samples, but using different samples is also plausible. RNA sample sources particularly need to be considered carefully, as the expression of many genes is restricted to certain cell or tissue types or upon a particular stimulation. If the hypothesis is related to gene activity on, for example, tumor cells, then a tumor specimen may be required.

There are three strategies to complete constructs A, B and C:

- *Strategy 1*: find the subset of genes that have both DNA variants and RNA expressions associated to the phenotype (i.e. constructs B and C established). Check for association between DNA variants and RNA expressions (construct A).
- *Strategy 2*: similar to strategy 1 but in a more relaxed way. Genes adjacent to associated DNA variants are included in the analysis. Similarly, genes in the upstream or downstream pathways of associated RNA expressions are considered. The subset of genes is then calculated by the intersection of the augmented gene sets.
- *Strategy 3*: use the conditional probability to depict the causal model as

  $$\Pr(L,R,C) = \Pr(L)\Pr(R|L)\Pr(C|R)$$

  where a DNA variant on a locus is denoted as L, the clinical phenotype is denoted as C, and the RNA expression is denoted as R (Schadt et al., 2005). The probability of clinical phenotype is conditioned by the DNA variant and RNA expression.

The construct A in strategy 3 is denoted as PR(L|R), which can be established by the analysis of quantitative trait locus (QTL), used to refer to the connection between the DNA variants to quantitative traits, such as gene expressions, or quantitative diseased states, such as elevated fasting glucose levels and HBa1C measurements in diabetic patients. The expression QTL (eQTL) is particularly used to refer QTL analysis correlating DNA variants to its corresponding RNA expressions.

Published by Woodhead Publishing Limited, 2013

Naturally, the analysis can be extended to protein level analysis to establish a more solid mechanistic explanation such as:

$$Pr(L, R, P, C) = Pr(L)Pr(R|L)Pr(P|R)Pr(C|P)$$

where the protein expression is denoted by P.

Despite the analysis that causal inference can eliminate several false positives, and increase evidence level, it remains a difficult task to find driver somatic mutations in cancer studies. This is because cancer has a complex progressive development. So many somatic mutations occur during the progress. The drivers are assumed to be critical in the process. The passengers could also demonstrate a mechanistic link of constructs A, and weakly B and C, and they may not play essential roles.

3.7 RNA secondary structure analysis

It is becoming clear that RNA molecules do not merely serve as passive information templates, but also carry important enzymatic and regulatory functions. RNA transcripts have secondary structures that are related to the half life, the stability, the functions and the interactions with other macromolecules. Computational analysis of RNA secondary structure is thus important for characterizing its roles. RNA secondary structures of one or two molecules can be predicted using the standalone Vienna RNA package (Hofacker et al., 1994; Hofacker, 2003). This is mainly based on the thermodynamic properties of base pairing, the net energy of which can be estimated as kilocalories per mole of RNA (kcal/mol).

3.8 Case studies

3.8.1 Molecular signature for allograft rejection after renal transplantation

Kidney transplantation is one of the few therapeutic choices currently available for patients with end-stage renal diseases. For such a therapy, recipients' acute immune rejection to allografts has been the major cause of failure. The detailed molecular mechanism of allograft rejection is still elusive. Furthermore, the histology of biopsy samples does not offer sufficient information for prediction of allograft rejection (Sarwal et al., 2003). The exploration of allograft transcriptomes may shed light on this critical yet unclear mechanism.

Sarwal et al. (2003) reported a gene expression study on biopsy specimens of patients with acute rejection after renal transplantation. A total of 67 biopsy samples, from 59 children and young adults, and 8 donors, were examined. Among them, 9 had allograft loss (which means treatment failure) due to intolerable host rejection. A complementary DNA (cDNA) microarray was used for the exploration mechanism, where the expression levels of 14,220 genes were measured using 28,032 probes. All of the samples showed no signs of lymphoproliferative disorder by histological examination which, therefore, cannot distinguish patients with elevated risks of allograft loss from the others.

The authors conducted a two-stage exploration analysis. The first stage was a grouping of samples using the bottom-up hierarchical clustering method. Gene expression levels were first normalized as folds with respect to the gene-specific mean expressions across all samples. This ensured the mean values of every single gene to be 1, therefore genes with low expression levels could also contribute to the grouping of samples. The similarity between pairs of samples was then quantified by Pearson correlation coefficients. The clustering was done as if the clinical status of the subjects was unknown, that is, an unsupervised method. The samples were grouped by the algorithm into four major clusters:

A later called the allograft rejection 1;

B comprising two groups, allograft rejection 2 and toxic drug effects and infection;

C also comprising two groups, allograft rejection 3 and chronic allograft nephropathy; and

D later confirmed to be the normal tissues.

An interactive heat map is presented online on the authors' website at Stanford University, using the GeneExplorer software to show the distinct characteristics of the four groups.

Based on the clusters defined in the first stage, the authors then examined the gene expression contrasts using the SAM method. They identified a focused set of 385 genes over-expressed in the allograft rejection 1 group compared with other subtypes. A functional annotation (based on enrichment analysis using the hypergeometric test) showed that the T cell related genes are particularly enriched in the over-expressed set of genes. This is consistent with the known significant role of T cell infiltration in

Published by Woodhead Publishing Limited, 2013

the rejection mechanism. What is unexpected is the appearance of several B cell related genes, particularly the CD20, on the focused gene set. The roles of CD20 and B cell infiltration are the major novel discovery in kidney allograft rejection. All the methods used (hierarchical clustering, SAM, enrichment analysis) have been detailed in previous sections.

To confirm the findings of RNA level microarray analysis, a monoclonal antibody of CD20 was employed for the immunostaining on 20 biopsy samples, a subset of the original study cohort. It was found that 8 out of 9 CD20 positive samples belonged to patients with allograft loss or incomplete functional recovery. In contrast, only 1 out of 11 CD20 negative samples was associated to such a poor condition ($P<0.001$). The protein level results further confirmed the findings on the RNA level exploration, a step toward clinical use. The corresponding Positive Predictive value was 88.9% and Negative Predictive Value was 90.9 (see Chapter 6 for definition). This is an excellent performance, as long as it can be validated in independent cohorts with larger sample sizes. In the future, immunostaining of CD20 might improve the clinical evaluation of the risks of allograft loss.

3.8.2 Molecular model for follicular lymphoma prognosis

Follicular lymphoma is a heterogeneous disease with patients' prognosis and life expectancy varying greatly, ranging between 1 to 20 years (Dave et al., 2004). The disease in some patients manifested a long-term dormant state, while in others it manifested a rapid progression. Causes for the variability remain a mystery. Thus, personally optimized treatment of follicular lymphoma has not yet been established. Exploration of the molecular signature may help to illuminate the disease mechanism, which might lead toward the prediction of prognosis and therefore the optimal personalized treatment.

Dave et al. (2004) collected 191 pretreatment tumor biopsy specimens and assigned them evenly into the training group ($n=95$) and the validation group ($n=96$). Gene expressions on these specimens were assayed using Affymetrix U133A and U133B microarrays. These subjects then received a variety of treatments or were followed without treatment. Overall survival was observed. An association between genes and overall survival time was evaluated in the training group by the Cox proportional hazard model. The gene by subject matrix was then visualized as two heat maps, where genes positively and negatively associated to survival time were separately presented. The gene expression levels were adjusted to folds with respect to the gene-specific medians across all samples. This ensured the median of every single gene to be 1.

A hierarchical clustering was then performed on genes. Ten gene clusters were identified where genes within a cluster had similar expression profiles across subjects, with Pearson correlation coefficient $r > 0.5$ between each other. The expression levels of genes in a cluster were then averaged to serve as a representative value for that cluster. In this way, the data were boiled down to a 10 variable signature per subject. An exhaustive enumeration of all pairwise combinations was performed, and a combination of two variables, called the immune response 1 and 2 respectively, had the strongest prediction performance of overall survival ($p < 0.001$ in both the training and validation groups). Elevated immune response 1 was associated to longer overall survival, while elevated immune response 2 was associated to shorter survival. An attempt was made to add one more variable to the two-variable signature, achieving a prediction $p < 0.001$ in the training cohort and $p = 0.003$ in the validation cohort. However, the third variable does not seem to contribute to the total prediction performance and so was dropped from the model.

The patients were then sub-grouped by the two-variable model into four quarters. The corresponding Keplan–Meier curves were used for visualization. The survival medians of the subgroups were 13.6, 11.1, 10.8 and 3.9 years, respectively. The last quarter of patients deserve specialized attention, as they represented a high-risk group with shorter survival.

3.8.3 Molecular similarity between stem and cancer cells

The cancer stem cell model is an insightful hypothesis of cancer initiation, progression and resistance to chemotherapy, which has a profound impact on cancer biology and treatment strategies. This model suggests that a class of stem cell like cancer cells are at the top of the hierarchy of heterogeneous cancer tissue. These cancer stem cells are responsible for driving the progression and, unlike other cancer cells, cannot be easily eliminated by chemotherapeutic agents. Cancers have been known to exhibit characteristics of reverse differentiation, such as the known Epithelial Mesenchymal Transition frequently observed in the metastatic stage of cancer progression. Gene expressions of de-differentiation markers have actually been used for the clinical staging of cancer.

Published by Woodhead Publishing Limited, 2013

Ben-Porath et al. (2008) conducted an analysis of transcriptomics of tumor specimens to explore the similar molecular signature between stem and cancer cells. The analysis is a sophisticated, two-pass enrichment analysis of gene expression data, with gene expression similarity between cancer and stem cells, where the genes are grouped into modules to facilitate the analysis. It combines enrichment analysis mentioned earlier with the grouping of related genes, clinical features and sample types. First, the gene signature of the human embryonic stem cell (hES) is defined based on previous studies. For example, genes over-expressed in more than five stem cell profiling studies were defined as a hES gene set. Second, genes over-expressed in individual cancer specimens were evaluated to see whether they were enriched in the pre-defined gene sets. In other words, they mapped real array data of cancer into stem cell classes. Third, they further conducted a second pass of enrichment calculation to see whether samples of the same type had similar enrichment patterns. This is a way to collapse a large gene-by-sample data matrix into a smaller, biologically more interpretable "gene class" by "sample group" matrix. They concluded that cancer cells do over-express genes that are normally associated with stem cells.

There are two important features for this type of analysis. First, enrichment analysis is used where genes are represented in either the "on" state (e.g. gene is over-expressed) or "off" state (e.g. gene is under-expressed). The quantitative measures of gene expressions need to be thresholded, either in absolute quantities or fold changes, to get either an "on" or "off" state. The genes with "on" states are then mapped to the gene classes of interest to see whether they are enriched. Second, the enrichment statistics are based on hypergeometric tests. The collection of sample types will determine the result of enrichment analysis.

3.9 Take home messages

- RNA molecules carry important regulatory and enzymatic functions, in addition to serving as templates for protein synthesis.
- Microarrays and NGS are two important transcriptomic platforms, each with its strengths.
- Contrast level analysis is mostly done by statistical tests.
- Module level, enrichment analysis can be done by gene ontology (GOEAST), Connectivity Map analysis, DAVID, GSEA, Enrichnet and PANTHER.

- Visualization methods such as the volcano plots can be used to present and analyze the results of multiple contrast criteria, particularly when the transcriptomic dataset represents more than two phenotypes.

3.10 References

Ben-Porath, I., Thomson, M.W., Carey, V.J., Ge, R., Bell, G.W., et al. (2008), An embryonic stem cell-like gene expression signature in poorly differentiated aggressive human tumors. *Nat Genet.*, **40(5)**: 499–507.

Bhattacharjee, A., Richards, W.G., Staunton, J., Li, C., Monti, S., et al. (2001) Classification of human lung carcinomas by mRNA expression profiling reveals distinct adenocarcinoma subclasses. *Proc. Natl. Acad. Sci. USA*, **98**: 13790–5

Bolstad, B.M., Irizarry, R.A., Astrand, M., et al. (2003) A comparison of normalization methods for high density oligonucleotide array data based on bias and variance. *Bioinformatics*, **19(2)**: 185–93.

Carthew, R.W. and Sontheimer, E.J. (2009) Origins and mechanisms of miRNAs and siRNAs. *Cell*, **136(4)**: 642–55.

Cech, T.R. (2011) The RNA Worlds in Context. Source: Department of Chemistry and Biochemistry, University of Colorado, Boulder, Colorado 80309–0215. Cold Spring Harb. Perspect. Biol. 16 February. p. ii: cshperspect.a006742v1. doi: 10.1101/cshperspect.a006742.

Chaussabel, D., Quinn, C. and Shen, J. (2008) A modular analysis framework for blood genomics studies: Application to systemic lupus erythematosus. *Immunity*, **29(1)**: 150–64.

Dave, S.S., Wright, G., Tan, B., Rosenwald, A., et al. (2004) Prediction of survival in follicular lymphoma based on molecular features of tumor-infiltrating immune cells. *N. Engl. J. Med.*, 351(21): 2159–69.

de Hoon, M.J., Imoto, S., Nolan, J., et al. (2004) Open source clustering software. *Bioinformatics*, **20(9)**: 1453–4.

Dysvik, B., Jonassen, I. (2001) J-Express: Exploring gene expression data using JAVA. *Bioinformatics*, 2001 17(4): 369–70.

Eddy, S.R. (2001) Non-coding RNA genes and the modern RNA world. *Nat. Rev. Genet.*, 2(12): 919–29

Eisen, M. B., Spellman, P. T., Brown, P. O. and Botstein, D. (1998) Cluster analysis and display of genome-wide expression patterns. *Proc. Natl. Acad. Sci. USA*, **95**: 14863–8.

Gamsiz, E.D., Ouyang, Q., Schmidt, M., et al. (2012) Genome-wide transcriptome analysis in murine neural retina using high-throughput RNA sequencing. *Genomics*, 99(1): 44–51.

Geiss, G.K., Bumgarner, R.E., Birditt, B., Dahl, T., Dowidar, N., et al. (2008) Direct multiplexed measurement of gene expression with color-coded probe pairs. *Nat. Biotechnol.*, **26(3)**: 317–25.

Glaab, E., Baudot, A., Krasnogor, N., Schneider, R. and Valencia, A. (2012) EnrichNet: Network-based gene set enrichment analysis. *Bioinformatics*, **28(18)**: i451–7.

Published by Woodhead Publishing Limited, 2013

Golub, T.R., Slonim, D.K., Tamayo, P., Huard, C., Gaasenbeek, M., et al. (1999) Molecular classification of cancer: Class discovery and class prediction by gene expression monitoring. *Science*, 286: 531–7.

Gorlov, I.P. (2010) *GWAS Meets Microarray: Are the Results of Genome-Wide Association Studies and Gene-Expression Profiling Consistent? Prostate Cancer as an Example*. PLOS one

Hofacker, I.L., et al. (1994) Fast folding and comparison of RNA secondary structures. *Monatsh. Chem.*, **125**: 167–88

Hofacker, I.L. (2003) Vienna RNA secondary structure server. *Nucleic Acids Res.*, **31(13)**: 3429–31.

IHGSC (2004) Finishing the euchromatic sequence of the human genome. *Nature*, **431**: 931–45.

Ideker, T., Dutkowski, J. and Leroy Hood, L. (2011) Boosting signal-to-noise in complex biology: Prior knowledge is power. *Cell*, **144**: 860–3.

Irizarry, R.A., Bolstad, B.M., Collin, F., et al. (2003a) Summaries of Affymetrix GeneChip probe level data. *Nucleic Acids Res.*, **31(4)**: e15.

Irizarry, R.A., Hobbs, B., Collin, F., et al. (2003b) Exploration, normalization, and summaries of high density oligonucleotide array probe level data. *Biostatistics*, **4(2)**: 249–64.

Kerr, K.F. (2007) Extended analysis of benchmark datasets for Agilent two-color microarrays, *BMC Bioinformatics*, **8**: 371.

Kim, C.S., Riikonen, P. and Salakoski, T. (2008) Detecting biological associations between genes based on the theory of phase synchronization. *Biosystems*, **92(2)**: 99–113.

Kotz, K.T., Xiao, W. and Miller-Graziano, C. (2010) Clinical microfluidics for neutrophil genomics and proteomics. *Nat Med.* **16(9)**: 1042–7.

Lamb, J., Crawford, E.D. and Peck D. (2006) The Connectivity Map: Using gene-expression signatures to connect small molecules, genes, and disease. *Science*, **313(5795)**: 1929–35.

Mann, H.B. and Whitney, D.R. (1947) On a test of whether one or to random variables is stochastically larger than the other. *Annals of Math. Stat.*, **18**: 50–60.

MAQC (2006) The microarray quality control (MAQC) project shows inter- and intraplatform reproducibility of gene expression measurements. *Nat Biotechnol.*, **24**: 1151–61.

Mootha, V.K., Lindgren, C.M., Eriksson, K.F., et al. (2003) PGC-1alpha-responsive genes involved in oxidative phosphorylation are coordinately downregulated in human diabetes. *Nat Genet.*, **34(3)**: 267–73.

Nagano, T. and Fraser, P. (2011) No-nonsense functions for long non-coding RNAs. *Cell*, **145**: 178–81.

Okazaki, Y. et al. (2002) Analysis of the mouse transcriptome based on functional annotation of 60,770 full-length cDNAs. *Nature*, **420(6915)**: 563–73.

Patterson, T.A., Lobenhofer, E.K., Fulmer-Smentek, S.B., Collins, P.J., Chu, T., et al. (2006) Performance comparison of one-color and two-color platforms within the Microarray Quality Control (MAQC) project.

Peer, D. and Hacohen, N. (2011) Principles and strategies for developing network models in cancer. *Cell*, **144**: 864–873.

Qin, L. et al. (2004) Empirical evaluation of data transformations and ranking statistics for microarray analysis. *Nucleic Acids Res.*, **32**: 5471–9.

Rosner, B. (2006) *Fundamentals of Biostatistics*. Thomson, USA.
Saldanha, A.J. (2004) JAVA Treeview: Extensible visualization of microarray data. *Bioinformatics*, **20(17)**: 3246–8.
Sarwal, M. (2003) Molecular heterogeneity in acute renal allograft rejection identified by DNA microarray profiling, *New Eng. J. Med.*, **349**: 125–38
Schadt, E.E., et al. (2005) An integrative genomics approach to infer causal associations between gene expression and disease. *Nat Genet.*, **37**: 710–17.
Subramanian, A., Tamayo, P., Mootha, V.K., et al. (2005) Gene set enrichment analysis: A knowledge-based approach for interpreting genome-wide expression profiles. *Proc. Natl. Acad. Sci. USA*, **102(43)**: 15545–50.
Thomas, P.D., Kejariwal, A., Campbell, M.J., et al. (2003) PANTHER: a browsable database of gene products organized by biological function, using curated protein family and subfamily classification. *Nucleic Acids Res.*, **31(1)**: 334–41.
Tusher, V.G., Tibshiranim R. and Chu, G. (2001) Significance analysis of microarrays applied to ionizing radiation response. *Proc. Natl. Acad. Sci. USA*, **98**: 5116–21
WTCCC (2007) Genome-wide association study of 14,000 cases of 7 common diseases and 3000 shared controls. *Nature*, **447**: 661–78.
Zheng, Q. and Wang, X.J. (2008) GOEAST: A web-based software toolkit for gene ontology enrichment analysis. *Nucleic Acids Res.*, 36(Web Server issue): W358–63.

Published by Woodhead Publishing Limited, 2013

4

Proteomics

DOI: 10.1533/9781908818232.83

Abstract: Proteomics is the varieties, quantities, roles and dynamics of all proteins in a cell, tissue or organism. Proteins are structural or functional elements of cells, comprising sequences of amino acids assembled according to templates of DNA and RNAs. Their sequences determine their structure, thus their cellular functions. Post transcriptional modifications occur in most proteins. Protein dynamics result from synthesis and degradation, which are well controlled in normal physiology. They also work in groups, so concurrent expressions, localization and physical interactions can shed light on their cellular roles. Proteins can function in extracellular space, circulating via the blood stream to function far away from where they were generated. This is why many serum or urine proteins serve as clinical biomarkers. Many aspects of clinical relevance still wait to be discovered. Here we explore the key concepts, practical computational tools and research directions of proteomics. We show that *de novo* peptide sequencing is an inverse problem, which can be solved by the adequate use of assumptions.

Key words: epigenomics, mass spectrometry, motifs, protein domains, protein structures, protein interactions.

4.1 Introduction

Proteomics is the systematic identification and characterization of all proteins, their abundance in time and space, their structures, their cellular activities and their interactions with other macromolecules. Proteomics can refer to all proteins in a cell, a tissue or an organism. Proteins are macromolecules composed of linear chains of basic elements called amino acids. There are 20 different types of amino acids in the human body, which are assembled in sequence into proteins according to the instruction of their messenger RNA templates. This process is called translation, which is performed by complicated molecular machinery and takes place in the ribosomes of eukaryotic organisms.

A genetic code book defines the map of three consecutive ribonucleotide bases (codons or triplets) in the messenger RNA to the corresponding amino acid residues of the protein. A set of four ribonucleotide bases (A, U, G and C) can give rise to 64 different triplet combinations in total. Among them, the triplet "AUG" codes for the amino acid methionine, as well as the starting signal of translation (known as the start codon). Similarly, three other stop codons (UAG, UAA, UGA) indicate the termination of translation. Each of the other 60 codons dictates the assembly of 1 of 19 other amino acids. This is mediated by 22 different transfer RNAs (tRNAs), which have an anti codon to form covalent bond base pairing with the messenger RNAs. At the other end of the tRNA, a residue binding motif is in place to grasp the amino acid residue and add it to the newly synthesized proteins.

Proteins may reside in particular regions of the cell, such as the cell membrane and nucleus, or be secreted into the extracellular space to carry out their functions at a remote site. The proteome can signify not only the cell types but also the cellular states. The human genome project has located and substantiated 21,000 protein coding genes (IHGSC, 2004). It is now imperative to characterize these gene products in the form of proteins. As will be seen later, the DNA and RNA sequences are extremely valuable in many aspects of protein studies.

However, proteomics brings its own layer of complexity in addition to genomics and transcriptomics. The 21,000 protein coding DNAs represent more than 1,000,000 final products of proteins, including alternatively spliced RNAs and the extensive post translational modifications occurring in almost every single protein (Godovac-Zimmermann and Brown, 2001). Proteins are also known to be pleiotropic, meaning that they may carry out several distinct functions. The dynamic and variable expression profile of a protein within a cell cannot be estimated by RNA assays, due to the discrepancy between RNA and protein levels in time and space. This discrepancy is caused by time lags in translation, difference of half lives of RNA and protein, and various post transcriptional regulations of proteins.

The enzymatic or structural functions of proteins make them closer to phenotype than DNAs and RNAs. Samples of a proteomics investigation can be obtained from cultured cells, tissues or body fluids. Human body

Published by Woodhead Publishing Limited, 2013

fluids (e.g. urine and serum) are attractive sample sources of protein assays due to their ease of acquisition both in the exploratory stage and in clinical use. Consequently, many established clinical assays are based on urine or serum proteins.

Major directions of proteomics studies (and their core technology) are:

- Characterize the constituent proteins in a protein complex for illustrating unknown protein localizations and functions (protein identification).
- Characterize protein interactions between pathogen and hosts (protein interaction detection).
- Compare protein expression profiles across clinically distinct persons (quantitative proteomics).
- Compare protein expression profiles between different experimental conditions of a cell line (quantitative proteomics).
- Annotate protein sequence by functional domains (protein sequence analysis).
- Study protein structures for illustrating their functions (protein structure analysis).
- Epigenomics.

4.1.1 Epigenomics

Epigenomics, which literally means "on top of" genomics, is the investigation of heritable marks materialized by chemical modification of the nucleotide bases and histones. Histones are a major type of chromosomal proteins, which are compactly wrapped around by genomic DNAs to form chromatins, to reduce chromosomal volume and strengthen the structure. The chemical modifications alongside the DNA sequences carry heritable information, passed down from cells to their offspring, by the completion of either mitotic or meiotic cell cycles. Common epigenomic modifications include the methylation of the nucleotide cytosine, or the methylation, acetylation and phosphorylation of the histone proteins. All these modifications may affect the binding of transcription factors to the transcription elements, thereby changing the expression of genes.

Epigenomics often relies on proteomics platforms for the study of histone modifications. It has been reported that many diseases such as colon cancer are initiated by altered epigenetic modifications, and subsequently the altered expression patterns of genes such as HNPCC, MSH2 and MLH1.

4.2 Proteomics platforms at a glance

Proteomic studies can be classified as "targeted" or "non-targeted" investigations. A non-targeted proteome is more like the genome-wide approach, where all constituent proteins are analyzed or even quantified. However, this is a daunting task. In comparison, a targeted proteomic study aims to characterize and quantify a selection of proteins from a sample. The targeted proteins include the exploration candidates based on the study rationale (i.e. membrane proteins, mitochondrial proteins, phosphorylated proteins, etc.), or the validation candidates which were selected at the exploratory phase of the study. Although not a genome-wide investigation by strict definition, many investigators still refer to the targeted study as the "proteome".

Current protein biotechnology does not have amplification technology similar to the polymerase chain reaction (PCR) widely used for DNA and RNA assays. Hence, the detection limit of trace amounts of proteins determines the completeness of a proteomics study. Fortunately, the sensitivity of protein assays is improving rapidly and we can rely on more recent and sensitive methods for those interesting but less abundant proteins.

4.2.1 Antibody based assays

Antibody based assays are used for targeted protein investigations. An antibody is a naturally occurring immunological molecule which has a wide range of sensitivity and specificity in binding to an antigen (another protein or macromolecule). Multiple specificity has also been observed in many antibodies. In such cases, the antibodies can recognize multiple antigens with a similar epitope, which is the major site where the antigen interacts with the antibody. It is this specificity that enables the therapeutic antibodies to antagonize their corresponding antigens with minimal effect on other proteins. This specificity has also being used extensively by modern biotechnology. Proteins in cells or tissues can be identified by their binding with monoclonal or polyclonal antibodies, which are tagged by fluorescent or radiological signals. Immuno (Western) blotting and immunohistochemistry (IHC) are two typical antibody based assays. The

Published by Woodhead Publishing Limited, 2013

latter is particularly useful for in-situ detection of proteins. IHC and Western blotting are often performed on one or few proteins at a time and the throughput is relatively low. Therefore, they are widely used in mechanism investigation or validation rather than large-scale exploratory studies.

The enzyme linked immunosorbent assay (ELISA) is another antibody based protein assay, employing antibody immobilized on a solid surface to detect proteins in liquid samples. ELISA is often performed in batches. For example, 96 samples on a plate are processed concurrently. Thus, this is ideal for validation of findings. Furthermore, once a protein is substantiated in an ELISA validation, the same assay can be further developed as a highly specific clinical assay. The disadvantage of this assay is that antibodies can interact with each other, prohibiting the construction of multiplexed assays utilizing multiple antibodies concurrently.

The antibody is also the basis of immuno-precipitation (IP), a capturing method for a particular class of antigens such as proteins. For example, IP has been used for systematic capturing of protein tyrosine phosphatases and their oxidized forms (Karisch et al., 2011). IP can also be used to detect the macromolecules, which stably bind to the targeted protein, called co-immunoprecipitation (CoIP). If the targeted protein is a transcriptional factor, then the DNA bond to the target may indicate a transcription element. IP is also useful for epigenomic investigations, based on the IP of chromatins (ChIP), then followed by a high throughput technique such as next-generation sequencing (ChIP-Seq) and microarray (ChIP on Chip) for the detection or quantification of DNA fragments bound to the chromatins. It can also be followed by mass spectrometry (MS) for the detection of chromatin modifications.

4.2.2 Mass spectrometry based platforms

Mass spectrometry (MS) based platforms are the only type of platforms so far that have the potential to practically detect and quantify known and unknown proteins on a large scale, in other words, the non-targeted proteome. MS-based platforms comprise two connected stations: separation (also called fractionation or purification) and peptide identification (Godovac-Zimmermann and Brown, 2001). Under this common framework, a wide variety of instrumentation has been installed and experimental protocols implemented. The particular type of instrument and protocol will inevitably affect the optimum platform level analysis. Clinical samples are usually in the liquid phase with a complex protein mixture, hence the first step is to separate highly complex protein mixtures into more homogeneous fractions before protein identification can take place. This is done via 2-D gel electrophoresis (2-D gel) or high performance liquid chromatography (HPLC). Proteins are first denatured by sodium dodecyl sulfate (SDS) so that they can move in the gel more freely. The 2-D gel technique can disperse the protein components in a mixture, relocating proteins by isoelectric points in one dimension, and the molecular weight in the other perpendicular dimension. The proteins are then visualized on the gel surface using staining methods, and indexed by their coordinates. The protein fractions of interest are then digested and excised from the gel for further analysis.

The second step is the identification and characterization of fractionated protein mixtures by MS. Due to the limitation of molecular size in an MS, proteins are first digested (cleaved) by proteolytic enzymes into smaller peptide fragments. For example, the trypsin enzyme cleaves all the composition proteins at the sites of two amino acids, Lysine (K) and Arginine (R). The digested samples are then vaporized from the liquid phase into a gas phase by ionization or vaporization technology, for example, electrospray ionization (ESI) and matrix assisted laser desorption ionization (MALDI).

The vaporized peptide ions are then sent through an electric field in a chamber with a detector on the other side. The time taken for each peptide fragment to fly through the chamber and hit the detector indicates its molecular weight per charge (i.e. the mass charge ratio, m/z). By recording the time of flight and intensity of fragments, a peptide mass fingerprint (PMF) is produced which can be visualized as a spectrum. Typically, a PMF spectrum can be generated per sample fraction. The horizontal axis of the mass spectrum indicates the time of arrival of each peptide fragment, which then translates into the mass to charge ratio. The current resolution of mass is between 0.01 and 0.001 Dalton (Mann and Kelleher, 2008). The vertical axis is the intensity, which may indicate the abundance but with a lower resolution.

Tandem mass spectrometry (MS/MS) is an essential technique for a finer characterization of peptide components within the sample of interest, particularly when the protein cannot be identified by one-pass MS. The peptides first pass through a regular MS chamber to produce the first mass spectrum. Selected peptides are sent through the second chamber, and smashed into ions by the collisional force with the inert gas in the chamber. The ions are then guided through the second electric field. The time of arrival then defines the ion level MS/MS spectrum.

Published by Woodhead Publishing Limited, 2013

The separation of protein mixture into fractions aims to reduce the analytical complexity of each fraction to a practical, manageable level. Ultimately, we need to stitch together data from different fractions into an integral whole so as to obtain a non-targeted proteomics scale dataset. This relies on a well organized automation and computer infrastructure to control experiment logistics and reduce human effort and error. The combination of liquid chromatography (LC) and ESI MS is the most (semi-) automatic way of high throughput proteomics.

4.2.3 Protein interaction networks

Three platforms are available for a systematic assay of protein interaction networks: the yeast two hybrid (Y2H), pull down or coimmunoprecipitation. The Y2H systems are reporter systems showing that if two proteins can physically interact, they will form a transcription factor in the reported system to activate the expression of reporter genes. The pull down assay employs baits (the protein of interest labeled with a tag for the following affinity purification step) to fish for their interacting proteins called preys. The preys are then analyzed by MS, identifying all the interacting proteins in a complex (Ho et al., 2002). Coimmunoprecipitation can capture a protein by antibodies, and the proteins which interact with the target protein then can also be captured for the following MS analysis.

4.3 Protein identification by MS based proteomics

MS-based proteomics platforms can generate a multi-dimensional raw signal. LC-based sample separation methods give the first dimension of eluting time, which characterizes the physical property of the subfraction of the sample. However, the 2-D gel approach gives the first and second dimension. Peptide identification, as produced by ESI, generates a spectrum per fraction, offering one dimension of m/z and another dimension of signal intensity, an index of peptide abundance. An even higher dimension of data will be produced in an MS/MS platform. The multi-dimensional data is often noisy and not aligned properly across experiments. Noise comes from, for example, the alternating current power source which generates an unwanted low frequency signal component, or the varying detector status which generates unwanted signal variation. For a fair comparison of MS or MS/MS in different runs or batches, platform level analyses are required, depending on the data quality.

Background subtraction and smoothing are common methods for removing systematic error and underlying drifting effects in a spectrum. The eluting time needs to be aligned and data normalization may be required to compensate for the batch difference of signal strengths. A large collection of analysis algorithms and software has been proposed to deal with platform level issues throughout the process. Each has its distinctive strengths (Box 4.1). What is introduced in this chapter is a general framework on the major concepts. For a comprehensive review, see Listgarten and Emili (2005).

Box 4.1 Useful bioinformatics tools and resources for proteomics

A. Resources for protein annotations

- Human Protein Atlas

The Human Protein Atlas is an important resource to find the evidence of protein expressions in tissues, which is mainly detected by immuno histological staining.

- Pfam

Pfam is a web service hosted by the Wellcome Trust Sanger Institute. It offers protein families, and families of families called clans, analyzed by multiple sequence alignment and hidden Markov models. It also offers biological domain annotations for proteins. Gene symbols can be used to search for a protein of interest. Users can easily find all the proteins that share the same domain.

- PROSITE

This is a web service hosted by the ExPASy site. It offers patterns, in the format of consensus and variable regions, of a highly conserved part of protein sequences called motifs.

Published by Woodhead Publishing Limited, 2013

- SCOP

SCOP stands for structural classification of proteins. It is hosted by Cambridge University. As the name suggests, the site offers classification based on 3-D similarity. It also offers convenient links to the structure image in PDB.

- PDB

PDB contains comprehensive protein structures detected by NMR or X-ray crystallography.

- SWISS-Model

The SWISS-Model offers the 3-D structure estimation and folding of proteins based on their sequences, using a template of known structures documented in PDB. This is a homology-based approach where the sequence homology is first computed, then the unknown structure is computed by the adaptation of known protein structures of the homologous proteins.

- InterPro

A single portal site, hosted in EBI, of many protein annotation databases such as Pfam and Prosite, InterPro can be used to find ortholog and paralog sequences of a gene easily.

- PANTHER

PANTHER was introduced in Box 3.1 to serve for module level enrichment based annotations on transcriptomics data. It is worth noting that this site also offers valuable protein classifications based on protein sequences.

B. Software for mass spectrometry based proteomics

- MaxQuant

MaxQuant is a package of software designed for quantitative analysis of the high resolution mass spectrum. It was developed by members of the Max Planck Institute of Biochemisty. The package offers a total solution, by integrating external ion matching software such as Mascot or Andromeda.

- Mascot

Mascot is a commercial package for PMF and MS/MS ion matching. Being a pioneer in this type of software, it is widely known and used.

- Flicker

Flicker is a 2-D gel image comparison software. It is hosted by NCI and also has a downloadable version.

- MRMer

MRMer is dedicated to MRM analysis. It has a free website service hosted by Fred Hutchinson Cancer Research Center (FHCRC). It also offers free JAVA software download (Martin et al., 2008).

4.3.1 Feature extraction

Peptide identification and comparison rely on accurate m/z and occasionally abundance measurements of peptides, shown as peaks in the spectrum. The feature extraction step is to compute accurate readings from the multi-dimensional analog and noisy data, with high sensitivity, specificity and resolution. It is commonly done by fitting a Gaussian envelope on the m/z intensity plane, then grouping the envelopes along the retention time axis with other layers of signal smoothing, to obtain smooth Gaussian hills. The peak of an envelope hill is then determined to give an m/z measurement and an intensity. All the peaks associated to a peptide are used to calculate the peptide abundance, which is an average of the intensities.

Despite the fitting and smoothing processes which have grouped fragmented signals into smooth hills, large amounts of noise still remain, which give false positives. A series of filters and grouping methods are then applied. One filter is by the m/z gaps between peaks in light of the legitimate weight ladder of amino acids. Regarding the MS/MS data, the weights of a-, b- and c-ions are related by their chemical formula. Using this information, we can group these ions so as to produce a much clearer spectrum.

Isotope pattern are also processed and are valuable cues. Isotope peaks are extracted purposely if a quantitative comparison of different protein sets is prepared with different isotopes. If the focus of the project is on the

Published by Woodhead Publishing Limited, 2013

identification of novel proteins, then the isotope peaks are grouped together.

The feature extraction can be performed by the MaxQuant package (Cox and Mann, 2008) (Box 4.1).

4.3.2 Computational spectrum matching

The identification of constituent proteins from samples such as cell lysates, body fluids or tissue extracts, is one of the major tasks in proteomics studies. A series of mass spectra may be produced for a sample which contains complex protein mixture. The way to interpret a spectrum is to match the m/z patterns of the spectrum computationally with those of known peptides called reference spectrums. If no match is found, the spectrum can also be matched against the theoretically predicted peptide m/z derived from genomic and transcriptomic sequences. This matching approach is called PMF. Proteins are identified if complete or partial matches are found. A similar computational approach can be applied to MS/MS spectrum analysis, such as the MS/MS ion matching method or the sequence tag method. The success of both the PMF and MS/MS ion matching relies on the completeness of the reference spectrums.

Several PMF and MS/MS ion matching software packages are available, such as Andromeda and Mascot (Box 4.1). Ideal computational matching methods need to be robust to accommodate noise and uncertainty. Typical source of uncertainties include:

- novel proteins;
- noise in spectra;
- unwanted cutting sites by other endogenous enzymes;
- unexpected post translational modifications.

These uncertainties remain challenging for protein (peptide) identification by computational spectrum matching, particularly when the noise and uncertainties may be instrument dependent. The *de novo* peptide sequencing is an alternative for MS/MS ion matching.

4.3.3 De novo peptide sequencing using tandem mass spectrometry

The *de novo* peptide sequencing methods aim to derive the sequence of the peptides directly from the MS/MS spectra, without using computational matching. This is the approach used to identify novel proteins, which does not exist in current databases.

An MS/MS spectrum has a very complex pattern due to a variety of ion types after the collision. The backbone of a protein molecule starts with a nitrogen atom (the N terminal) and finishes with a carbon atom (the C terminal). An amino acid can be represented by a unit of three atoms (N, C, C). Upon collision, a peptide can break down in many different places in the molecule, particularly along the backbone and between the boundary of two amino acids, (i.e. the C–N bond). In such a case, the parts ahead of the break are called b-ions, and the parts behind the break are called y-ions (Figure 4.1). The break might also occur in other places. In such cases, the parts before and after a break of the (C–C) bond are called a- and x-ions, respectively. The parts before and after the break of the (N–C) bond are called c- and z-ions, respectively (Figure 4.1).

The *de novo* peptide sequencing is a typical inverse problem. The corresponding direct problem is to generate a hypothetical spectrum for a peptide sequence (Figure 4.2).

A *de novo* peptide sequencing algorithm may be derived based on the practical tool of graphs, which serve as abstract representations

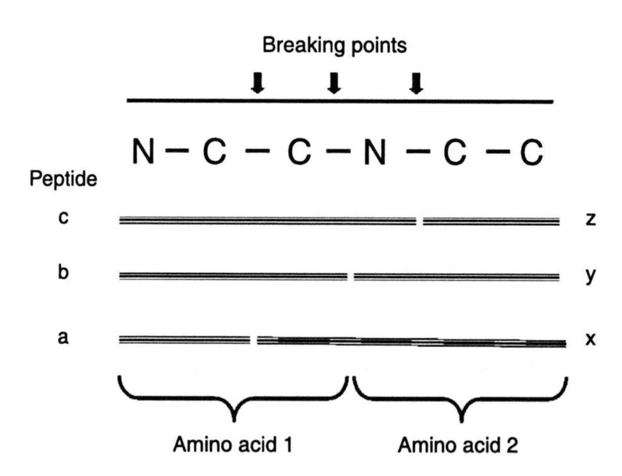

Figure 4.1 The breaking points result in different types of peptide fragments after collision. The break not only occurs between the peptide bonds, producing b- and y-ions, but also occurs within an amino acid residue and along the backbone, producing x- and z-ions from the C terminal (called C terminal ions)

Published by Woodhead Publishing Limited, 2013

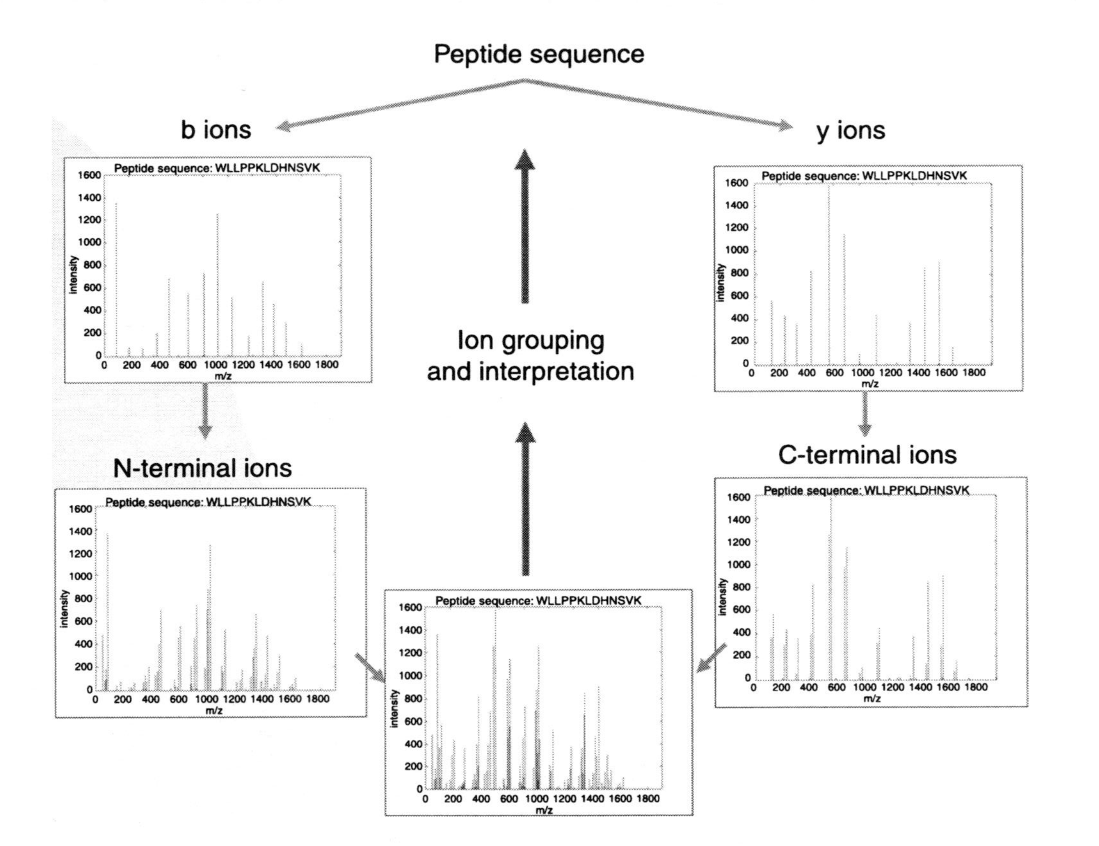

Figure 4.2 The direct and inverse process from peptide to spectrum. The downward arrows represent the route or direct process to generate a theoretical spectrum from a peptide. The upward arrows represent the inverse process for grouping and analysing signals in the spectrum to generate the sequence

Published by Woodhead Publishing Limited, 2013

of MS/MS spectra. The graph is often defined by the node (also called the vertex) and link (also called the edge) nomenclature (Barabási and Oltvai, 2004). Graphs have two types, directed and undirected, depending on whether the edges have directions. For example, protein interaction networks are often illustrated as undirected graphs, due to the reciprocal relationship between the interacting proteins, while transcriptional networks or signal transduction networks are often modeled as directed graphs. Here we will use the directed graph for *de novo* peptide sequencing. The link could be weighted based on the original data. The weight is often inversely represented by the length of the link. In this algorithm, the spectrum is first preprocessed and transformed into a graph, then we devise a graph traversing algorithm so as to obtain the sequence.

Preprocessing of the spectrum for *de novo* sequencing

The main purpose of spectrum preprocessing and ions grouping is to clean up the spectrum to contain only b- and y-ions and without peaks by partial amino acids, for the ease of amino acid level analysis. The spectrum preprocessing includes the technique of spectrum filtering and isotope. The ion grouping step associates a-, b- and c-ions together as b-ions, and x-, y- and z-ions as y-ions.

Graph construction

To characterize a spectrum, we need three basic elements for the graph: node, directed link and undirected link. Nodes are used to represent ion peaks. The number of nodes in the spectrum graph is $k + 2$, where k is the number of peaks in the spectrum and the two additional nodes are for 0 and the precursor. The undirected link represents the relationship between b and y pairs. The idea is to employ the known molecular weight of the precursor peptide to associate b- and y-ions as pairs. By subtracting the molecular weight of a b-ion from the precursor weight, we obtain the weight of the accompanying y-ion. Finally, directed links are used to connect peaks toward another peak which has an additional weight of one amino acid. The graph is generated in three steps. First, create nodes for every single peak of b- and y-ions. Second, connect nodes of accompanying b- and y-ions by undirected links. Third, connect nodes which have a weight difference of one amino acid by directed links.

Published by Woodhead Publishing Limited, 2013

Graph traversing

A graph walking algorithm is then in place to read out the peptide sequence from this particular type of graph. Starting from the node with smallest molecular weight, we work through the directed link (i.e. increasing molecular weight) one after another until no directed link remains. If that happens, then the last undirected link encountered will be traversed instead. After passing the undirected link, work will continue on directed links but with a change of direction (i.e. turn to the inverse direction, which means a decrease in molecular weight). The work will continue until it cannot go any further. Then the last encountered undirected link is passed again and the traverse continues with increasing mass. This process is continued until the reading of the sequence is complete.

4.3.4 Quantitative proteomics

Quantitative measurements of abundance (in terms of concentration, molecule number (mole), or total molecular weight) of multiple peptides across different experimental conditions are an important goal of proteomic investigations or quantitative proteomics. However, this depends on whether the technology can provide accurate quantitative estimates across all samples. The stable isotope labeling by amino acids in cell culture (SILAC) method was designed to fulfill *in vitro* or *ex vivo* quantitative proteomics (Ong et al., 2002). It is an experiment protocol designed to evaluate paired cell-line samples under two experimental conditions. The paired samples are also cultured with and without an isotope medium, respectively. This is because a variety of noise and bias could occur during both the separation and identification steps of the proteomics process. Hence, a pair of samples are pooled together to go through the same MS process, cancelling out most of the bias. However, the application of this method is confined to cell-based samples, such as *in vitro* or *ex vivo* samples, due to the requirement of cell culture. It cannot be used to assay clinical samples directly. For such studies, methods such as isobaric tags for relative and absolute quantification (iTRAQ) can be used.

The recent multiple reaction monitoring (MRM) technique offers reliable MS-based quantitative measurement of clinical samples (Chen et al., 2012). Absolute quantification is achieved by a calibration step in MRM. This technology is particularly useful when the ELISA assay (another technique adequate for quantitative proteomics) is not available, or when multiple proteins are targeted so that MRM offers a multiplexing solution. A proteomic data set, either of binary or quantitative values, can be organized as a peptide by sample matrix, similar to the transcriptomic gene by sample matrix. As such, many contrast level analysis methods for transcriptomics studies can also be used for proteomics data analysis, as long as the underlying assumptions are met, for example, Mann–Whitney tests may be more suitable than non-paired t-tests in MRM data, because peptide concentration across samples may not fit a normal distribution.

4.4 From protein sequences to structures

4.4.1 Protein motifs and domains

Protein sequences dictate their 3-D structures, which in turn determine their functions. Proteins can be organized in a hierarchy of superfamily, family and subfamily, based on their resemblance to structures and functions. From the evolutionary perspective, member proteins of a family are derived from ortholog and paralog events. The former means one protein is the ancestor of the other. The latter means two genes are produced in a genomic duplication event during evolution. The annotation of novel proteins into families is mainly based on their composition domains, which are conserved regions of a family with specific folding structures and functions. Motifs are conserved amino acid sequences of proteins. The conservation of motifs in evolution implies their importance in structures and functions.

The homology of protein sequences can be analyzed by the following methods. First, a multiple sequence alignment can be used to align the residuals of related protein sequences. Second, a position specific scoring matrix can be constructed. Highly conserved residuals can be identified as motifs. Highly conserved regions are annotated as domains. The sequence signature of a protein family can be represented using regular expressions, or hidden Markov models. They can be used to check whether a newly discovered protein belongs to this protein family.

Protein domain information can be found in Pfam, while motif can be found in Prosite. Protein structure information can be found in PDB and SCOP (Box 4.1). All these resources have been integrated into a single portal called InterPro. Using the graphical presentation of InterPro, it is

Published by Woodhead Publishing Limited, 2013

easy to see that the boundaries of protein domains annotated by different algorithms are slightly different.

4.4.2 Computational protein structure prediction

Proteins are the structural and functional components of a cell. The functions of proteins are largely determined by their 3-D structures (sometimes called shapes or conformations). A protein may have multiple conformations. They undergo conformational change to alter their functions. The structure has multiple layers of regularities. The primary structure refers to the 1-D amino acid sequences. The secondary structure is the highly regular local substructure of the small segments of proteins known as motifs. Alpha helix and beta sheet are two common secondary structures. The tertiary structures refer to the 3-D structures of a single protein. The quaternary structures refer to the 3-D structures of protein complexes, which are important conformations, because proteins usually form complexes to fulfill their roles.

Protein structure can be experimentally determined by X-ray crystallography or nuclear magnetic resonance (NMR) spectrography. The former requires the purification and crystallization of proteins. The latter can only be used to analyze small molecules and also requires the purification of proteins. Many determined protein structures have been documented in a public domain database PDB (Box 4.1). The determination of a protein structure experimentally is challenging. For example, the structure of many membrane proteins cannot yet be determined.

Computational approaches for the determination of protein structures are thus very important to fulfill this unmet need. One approach is by *ab initio* protein folding, which means the estimation of 3-D structures from the "folding" of 1-D amino acid sequences. This is based on a global minimization of free energy considering the atomic interactions. In the interest of time and resource saving, local optimization may be implemented in practice and gradually converge to globally sub-optimal results.

A protein family is a collection of proteins with similar structures, which are evolutionarily conserved to carry out important protein functions. If the structure of a family member can be determined experimentally, the rest of the family members can be computationally folded by the use of known protein structures. Folding by known protein structures can dramatically save time, as the computation now addresses only the local minimization of energy.

When no known protein structure is available in a protein family, then computation become more challenging. Recently, software of EVfold_membrane has demonstrated excellent performance in the *de novo* folding of alpha helical transmembrane proteins without the use of known structures (Hopf et al., 2012). This software is based on the assumption that if two amino acid are physically bound and are of structural importance, then they should co-evolve together. By analyzing the amino acid sequence of a protein family in evolution, the co-evolved residuals can indicate the potential bonding, thereby guiding the computational folding of the structure efficiently.

The design of drugs to antagonize disease-related proteins requires their structural information. Computational software has been developed to simulate the interaction of small molecules into macromolecules such as proteins. This is called docking.

4.5 Protein interaction networks

Proteins exert their functions through physical interaction with each other. An enzymatic process is a typical example, where an enzyme makes physical contact with its substrate and then triggers a reaction. Many receptor proteins on the cell surface need to form a dimerized or trimerized (another type of interaction) structure before they can function properly and send an external signal to the cell. Proteins also work in groups or complexes. They interact with several other proteins, concurrently or sequentially, in order to accomplish their function. The studies of physical contacts among proteins can illuminate cellular roles of proteins.

Investigations on protein interaction networks so far have been restricted to simpler organisms such as the budding yeast (*Saccharomyces cerevisiae*). A study employed 725 baits, including 100 protein kinase, 36 phosphatases and regulatory subunits (as well as 86 proteins implicated in DNA damage response), covering approximately 10% of the predicted total yeast proteins (Ho et al., 2002), and 1578 preys were identified, comprising another 25% of the yeast proteome. The complexes were separately identified from different subcellular compartments, including the cytoplasm, cytoskeleton, nucleus, nucleolus, plasma membrane and mitochondria.

It should be noted that protein interactions vary in time and space. The natural barrier of cell compartmentation may prohibit certain pairs of proteins from interacting. But this natural barrier may be disrupted in some experiments of Y2H or CoIP, resulting in false positive interactions. One other problem for the Y2H assay is that it lacks time information. If

Published by Woodhead Publishing Limited, 2013

two proteins are always expressed at different times, they do not have the opportunity to bind, although they may give a positive signal in the artificial reporter system.

4.5.1 Filtering and integration

A complete characterization of all protein interactions in a species is challenging. The data may contain false positives, while at the same time being incomplete (von Mering et al., 2002). This requires data filtering and integration to improve data quality.

A filtering of interactions based on prior knowledge of protein compartmentation can reduce the number of false positives. Furthermore, the interaction of proteins in their native forms may vary in time, yet an experiment usually only captures a snapshot in a particular cellular state. Interactions occurring at different time points may be missed by the experiment. Considering the distinct strength and weakness of different technological platforms in terms of sensitivity and specificity, it is often useful to integrate results from different experiments and different platforms, rendering a more complete picture (von Mering et al., 2002).

4.5.2 Data presentation

Presentation of interaction data is critical to reveal underlying structures, and facilitate the reasoning and computation, as well as storage. It is imperative to present a global view of protein interactions in the context of networks in various cells, tissues or species under different conditions.

Protein–protein interaction can be represented as a 2-D matrix (Guruharsha et al., 2011). If we rank proteins in both the column and row of the matrix, then a dot in the matrix represents an interaction. Because the interaction between two randomly selected proteins is rated, this presentation will result in a sparse matrix. A sparse matrix representation is easy to implement but requires a huge amount of space, particularly when the number of proteins is large.

A graph with nodes and links can present a global view of a protein interaction network of an organism. Nodes represent proteins and links indicate the interaction. A graph or network is normally presented using either a link list or hyperlink data structure, which can carry the same message as a sparse matrix. A link list representation is economical in storage and also facilitates a web presentation of hyperlinks connecting web pages of protein which interact with each other.

After the network was constructed from experimental data on budding yeast (Ho et al., 2002), hubs were identified, which had more links than average to other proteins. A following series of gene knock out experiments showed that the hubs were essential proteins, which are critical for survival.

4.6 Case studies

4.6.1 *Interactions of viral proteins and human proteins*

Viral infections are known to cause a variety of human diseases, including cancer. Viral proteins are expected to elicit a spectrum of pathogenic events. It has been well accepted that cancer is a genetic disease where a spectrum of somatic mutations drives the initiation and progression of this disease. At the same time, viral proteins are known to interfere with the function of tumor suppressor genes such as TP53 and RB, thereby shutting down the endogeneous cancer prevention mechanisms. It was argued that a systematic evaluation of viral protein vs. human protein interactions can decipher additional carcinogenesis mechanism elicited by the viral protein (Rozenblatt-Rosen et al., 2012).

Rozenblatt-Rosen and colleagues conducted a systems level investigation on viral open reading frame (ORF) proteins of four different viruses, including human papillomavirus (HPV), Epstein–Barr virus (EBV), adenovirus (Ad5) and polyomavirus (PyV). The first step utilized the viral vs. human protein interactomes detected by Y2H experiments. Interactions were then validated by the tandem affinity purification MS on a human cell line expressing the viral protein ORFs.

The next step was an assay of transcriptomic alteration in response to the stimulation of viral proteins, which has been validated to interact with human proteins. Many cancer pathways are thus illustrated.

4.6.2 *Detecting proteins which selectively interact with epigenetically modified histones*

Epigenetic modifications of histone proteins may affect the binding of transcription factors, thereby altering RNA expressions. However, the

Published by Woodhead Publishing Limited, 2013

transcriptional factors, which show specificity to modified histones but not unmodified ones remain unclear. Characterizing these proteins is an important step toward a systematic understanding of epigenetic regulations.

Vermeulen et al. (2010) set out to investigate the proteins which interact only with modified histone proteins. Two expression activating modifications (H3K4me3 and H3K36me3) and three expression suppressing modifications (H3K9me3, H3K27me3 and H4K20me3) were investigated. They employed forward and reverse SILAC assays together with the chromatin immunoprecipitation (ChIP), by which the nuclear extracts were pulled down by modified (e.g. H3K4me3) and unmodified (e.g. H3K4) histones for comparison. HeLaS3 cells were cultured in heavy and light media. In the forward assay, cell lysates with the heavy label were pulled down by modified proteins, while those with the light label were pulled down by unmodified proteins. The opposite arrangement of label was also used in the reverse assay to balance out the different label efficiency. Proteins specifically interacting with modified histones were identified by a large ratio of heavy ions against the pairing light ions in the forward assay, as well as a small ratio in the reverse assay. MaxQuant was used for spectrum data analysis.

4.6.3 Identifying protein composition of mitotic chromosomes

Most chromosomal proteins involved in the mitotic stage of the cell cycle remain unknown, despite a handful of kinetochore genes which have been well characterized. It is a challenge to identify mitotic chromosomal proteins, due to difficulty in the isolation and purification of this sub-cellular organelle without contamination by other proteins. Mitochondrial proteins are a critical source of contamination. In addition, many "hitchhiker" proteins attach to the highly charged mitotic chromosomes (histones are positively charged and DNA are negatively charged) simply by electrostatic forces rather than actually playing a role. Given a high contamination level, the characterization of chromosomal proteins is not merely a protein identification task, but also a classification task aimed to tell apart true chromosomal proteins from false positives such as hitchhiker contaminants.

Ohta et al. (2010) employed five different types of proteomics experiments and one bionformatics analysis, termed six classifiers, for the joint classification of mitotic chromosomal proteins. Although each classifier alone cannot render a result with low contamination, a combination of six classifiers can. The six classifiers are:

1. abundance estimation;
2. enrichment in chromosomes;
3. *in vitro* exchange on chromosomes;
4. SMC2 dependency;
5. Ska3/Rama 1 dependency;
6. domain Analysis.

Classifiers 2 to 5 employed SILAC technology. These classifiers identify proteins that have a large contrast in light and heavily labeled proteins. Classifier 6 is a bioinformatics analysis counting the domains appearing in chromosomal proteins (but not the mitotic stage of proteins). Since each classifier gives a quantitative result for a protein (e.g. abundance measurement, light/heavy ratio), a multivariate classifier may be used to combine these quantitative results into a dichotomous decision (whether the protein is a mitotic chromosomal protein). The authors chose random forest classifier, decision tree based methods, which generate a combined score for each peptide for classification. Using the chicken DT40 cells as the material, 4000 proteins were identified as mitotic proteins in this way. MaxQuant software was used for platform level analysis, and Mascot was used for protein identification on MS and MS/MS spectra.

4.6.4 Protein codon order affects the translation efficiency

The translation of protein from RNA involves 61 different types of codons, which code for 20 different types of amino acids. Most tRNA associate with only one codon of an amino acid, with the exception that a few tRNAs can associate with two codons of one amino acid. Since a protein is usually composed of hundreds to thousands of amino acids, the same amino acid many appear multiple times in a protein sequence. It is thus an interesting question to ask whether nature prefers the use of the same tRNAs whenever an amino acid is required. To rephrase the question, is the same codon repeatedly used for the same amino acid?

Cannarozzi et al. (2010) analyzed the entire set of protein sequences in yeast. They analyzed all serines, which are encoded by 6 different codons with 4 different tRNAs. They counted the codon co-occurrence and produced a 6 (codon) by 6 (codon) contingency table. This can be

Published by Woodhead Publishing Limited, 2013

collapsed into a 4 (tRNA) by 4 (tRNA) contingency table, if we merge the counts of the same tRNA. Using a Chi-square test, we obtain very small P-values ($P < 0.0001$), showing that both contingency tables have data distribution deviating from the expected (random) distribution. This shows that the tRNA and codon usage is not random.

Cannarozzi et al. (2010) went one step further to demonstrate that nature prefers the same tRNA to be used repetitively. This can be shown if the counts are more concentrated on the diagonal cells of the 4×4 table. This concentration is more enhanced when the standard deviation from expectation of each cell is shown, where the diagonal cells show large positive values, while the others show negative values. Thus they concluded the tRNA is often reused, which is postulated to help the translation efficiency.

4.7 Take home messages

- Proteins are classified into families and superfamilies, which reflects their evolution history. Proteins in the same family have similar sequences of amino acids, functional domains and 3-D structure.
- The homology of protein sequences and structures is critical for characterizing unknown proteins.
- The bindings and interactions of proteins are also critical for characterizing unknown proteins.
- ELISA and MS-based protein assays are the major platforms for proteomics studies of biomedical topics.
- Bioinformatics of MS related platforms includes feature (peak) extraction, PMF, ion matching and *de novo* peptide sequencing.

4.8 References

Barabási, A.L. and Oltvai, Z.N. (2004) Network biology: Understanding the cell's functional organization. *Nat Rev Genet.*, **5(2)**: 101–13.

Cannarozzi, G., Schraudolph, N.N., Faty, M., et al. (2010) A role for codon order in translation dynamics. *Cell*, **141(2)**: 355–67.

Chen, Y.T., Chen, H.W., Domanski, D., Smith, D.S., Liang, K.H., et al. (2012) Multiplexed quantification of 63 proteins in human urine by multiple reaction monitoring-based mass spectrometry for discovery of potential bladder cancer biomarkers. *J Proteomics*, 75(12): 3529–45.

Cox, J. and Mann, M. (2008) MaxQuant enables high peptide identification rates, individualized ppb-range mass accuracies and proteome-wide protein quantification. *Nat. Biotechnol.*, **26(12)**: 1367–72.

Godovac-Zimmermann, J. and Brown, L.R. (2001) Perspectives for mass spectrometry and functional proteomics. *Mass Spectrom. Rev.*, **20(1)**: 1–57.

Guruharsha, K.G., Rual, J.F., Zhai, B., et al. (2011) A protein complex network of *Drosophila melanogaster*. *Cell*, **147(3)**: 690–703.

Ho, Y., Gruhler, A., Heilbut, A., Bader, G.D., Moore, L., et al. (2002) Systematic identification of protein complexes in *Saccharomyces cerevisiae* by mass spectrometry. *Nature*, **415(6868)**: 180–3.

Hopf, T.A., Colwell, L.J., Sheridan, R., Rost, B., Sander, C. and Marks, D.S. (2012) Three-dimensional structures of membrane proteins from genomic sequencing. *Cell*, **149(7)**: 1607–21.

IHGSC (2004) Finishing the euchromatic sequence of the human genome. *Nature*, **431**: 931–45

Karisch, R., Fernandez, M., Taylor, P., et al. (2011) Global proteomic assessment of the classical protein-tyrosine phosphatome and "Redoxome". *Cell*, **146(5)**: 826–40.

Listgarten, J. and Emili, A. (2005) Statistical and computational methods for comparative proteomic profiling using liquid chromatography-tandem mass spectrometry. *Mol. Cell Proteomics*, **4(4)**: 419–34.

Mann, M. and Kelleher, N.L. (2008) Precision proteomics: The case for high resolution and high mass accuracy. *Proc. Natl. Acad. Sci. USA*, **105(47)**: 18132–8.

Martin, D.B., Holzman, T., May, D., Peterson, A., Eastham, A., et al. (2008) MRMer, an interactive open source and cross-platform system for data extraction and visualization of multiple reaction monitoring experiments. *Mol. Cell Proteomics*, **7(11)**: 2270–8.

Ohta, S., et al. (2010) The protein composition of mitotic chromosomes determined using multiclassifier combinatorial proteomics. *Cell*, **142**: 810–21

Ong, S.E., Blagoev, B. and Kratchmarova, I. (2002) Stable isotope labeling by amino acids in cell culture, SILAC, as a simple and accurate approach to expression proteomics. *Mol. Cell Proteomics*, **1(5)**: 376–86.

Rozenblatt-Rosen, O., Deo, R.C., Padi, M., et al. (2012) Interpreting cancer genomes using systematic host network perturbations by tumor virus proteins. *Nature*, **487(7408)**: 491–5.

Vermeulen, M., Eberl, H.C., Matarese, F., et al. (2010) Quantitative interaction proteomics and genome-wide profiling of epigenetic histone marks and their readers. *Cell*, **142(6)**: 967–80.

von Mering, C., Krause, R., Snel, B., Cornell, M., Oliver, S.G., et al. (2002) Comparative assessment of large-scale data sets of protein–protein interactions. *Nature*, **417(6887)**: 399–403.

Published by Woodhead Publishing Limited, 2013

5

Systems biomedical science

DOI: 10.1533/9781908818232.107

Abstract: Cells seldom live in isolation. They constantly interact with ambient cells and the environment. Many conventional studies have focused on particular molecular activities taking place within cells, which are cultured in artificial environments, so the rich interaction between the cells and their natural environment is unexplored. To acquire a more complete picture, systematic investigation of all cellular activities in different environmental settings would be valuable. With three distinct omics (genomics, transcriptomics and proteomics), we will see how new insights into human biology can be approached by systematic hypothesis generation and panoramic, multi-scale data filtering, integration and modeling. Systems biomedical science is an iteration of bottom-up and top-down investigations. Individual systems are simulated jointly for their emergent collective behavior. Case studies include whole cell modeling, synchronization and phase locking models of cell cycles, essential genes for the normal operation of cell cycles, and the hourglass model for embryonic development in many species.

Key words: modeling, simulation, conceptual framework, emergence, synchronization, internal variables.

5.1 Introduction

Since the dawn of history, the continuing pursuit of medical knowledge has created modern medicine, comprising an arsenal of diagnostic and treatment methods against many diseases. Looking back, tremendous advances in medical technology have been made. Looking ahead, many illnesses still remain poorly treated or even untreatable; to name a few, the acquired immune deficiency syndrome (AIDS) and many late-stage cancers. The limitations of modern medicine reflect our still incomplete and fragmented understanding of human biology, physiology, pathology and pharmacology, as well as the interaction with resident and environment microbes. Extending the frontier of knowledge not only requires time, but also a new way of thinking, which is the main topic of this chapter.

Conventional studies have three major problems:

1. *Over simplification*: scientists often pursue simple systems with few types of molecules (or genes). Albeit existent, the effect of these molecules on the whole physical system may be trivial.
2. *Isolation*: many biomedical studies are conducted using cell lines or laboratory animals under strictly controlled experimental conditions. This may not represent real biological behaviors.
3. *Narrow range*: a human study such as the phase-three clinical trial may demonstrate the drug effect and safety in a fixed dose, on patients of a particular age range or ethnic origins. This is akin to examining a cross-section of the entire universe of possibility, or a tip of the iceberg. Conclusions obtained this way often cannot be generalized in a straightforward manner, unless all the other scenarios have been meticulously examined.

The history of biomedical discovery is thus analogous to a group of blind people trying to describe an elephant based on their limited contact experiences.

Complexity and non-linearity are two major hurdles hindering a full comprehension of human biology. Non-linearity means that the joint effect of two elements exceeds or deviates from the linear combination of individual effects in isolation. Real biological systems are often non-linear, with multiple layers of information tangled together in a complex way. We are more familiar with the linear, logical reasoning of cause and effect, which has shown its strengths in previous studies. Yet, simple cause and effect relationships only exist in small, isolated systems involving only a few types of molecules.

Thus, a major direction of future biomedical science is in viewing the larger picture and relaxing the restrictions posted by conventional thinking. We have seen an unprecedented opportunity to systematically explore the molecular basis of human biology by high throughput "omics" technologies. Naturally, deep data of multiple biological conditions contribute to the broad applicability of the derived empirical rules, with

Published by Woodhead Publishing Limited, 2013

one caveat that novel insights are buried in the deluge of data. To decrease the overload of information, adaptive bottom-up and top-down processes become inevitable. Allowing the data to speak for themselves, the non-linear connection of biological entities and phenotypes may gradually arise from seemingly chaotic data, therefore rendering novel insights.

Systems biomedical science (SBMS) represents a new way of investigation. Rather than focusing on oversimplified, isolated and narrow mechanisms, SBMS features adaptive exploration of complex issues from multiple angles. A panoramic view is enabled by a systematic interpretation of a broad spectrum of data. The goal is to reveal novel, terse, universal, high-quality and general rules for future empirical validation. This way, biomedical science can ascend to the next level. Genomics, transcriptomics and proteomics technology all have their individual strengths and limitations. In Chapters 2 to 4 we have explored how to conduct a joint analysis from multiple types of data (i.e. systems level analysis). This chapter will further formalize systems level analysis involving multiple systems.

5.1.1 Systems: Across conventional boundaries

The term “system” is prevailingly used in the field of engineering, referring to a set of connected and confined mechanical components which can operate together. A system usually has a defined boundary, with inputs and outputs from and to the environment. In SBMS, a system has two definitions:

1. a multifunctional biological entity; and
2. a collection of the same type of entities, which can interact with each other.

In such a framework, an iterative definition is allowed that one system may comprise other (sub) systems. This is suitable for the modeling of the human body, which comprises hierarchical layers of biological entities: the molecular, cellular, tissue, organ and physiological layers. A layer (which is a system by definition 2) has multiple composition entities (which are also systems by definition 1), each with complex interactions with other entities in the same layer. For example, a tissue (a layer) comprises many cells (biological entities). SBMS addresses a collection of individual systems interacting together. Composite entities have shared property as well as their individuality. For example, cells of the same type all have cell cycles, each with a slightly different frequency and phase. A layer is featured by the collective behavior of composite entities. The organization of layers enables the investigation to be first focused on one single layer, thereby reducing the difficulties in understanding the complex integral at once. For example, a phenotype (i.e. a disease) often manifests characteristics (symptoms) in multiple layers. Ultimately, SMBS requires working across conventional boundaries and layer definitions.

5.1.2 A two way approach of top-down plus adaptive bottom-up

We define SBMS as the activity and result of adaptive bottom-up and top-down investigations. Empirical data represent the bottom layer, offering the source material for reasoning. The bottom-up approach is to gradually stitch together pieces of heterogeneous data and composition entity models with increasing abstraction and generalization, until a global integral is achieved. It is important to let the data speak for themselves, with the help of adaptive computational models, and observe carefully the non-linear, collective effects emerging from the interaction of systems. This is nicely addressed by the classic essay "More is Different" by the physicist P.W. Anderson (1972).

Thus, a purely "automatic", bottom-up approach has three limitations:

1. It cannot make an extrapolation of data beyond its original scope, and an interpolation within the data resolution. A higher level "abstraction" and "generalization" requires human inference and domain knowledge.
2. The genomic data of various levels are usually very noisy within which the true biological signals are buried. Data filtering and integration are required to isolate the true picture from background noise (Ideker et al., 2011). Adequate filtering and integration depend on domain knowledge on the characteristics (strengths and limitations) of the data, as well as on the underlying biological effect.
3. Human biology is complex and the enumeration of molecular interactions in different time and place is infinite.

Hence, a high-level conceptual framework is critical to put the diverse data into perspective. As mentioned in Chapter 1, general or specific

Published by Woodhead Publishing Limited, 2013

challenges during biomedical investigations can often be ascribed to inverse problems, requiring top-down assumptions and conceptual frameworks to reach solutions.

5.2 Cell level technology and resources at a glance

Cell-level technologies are critical for deep phenotyping, an important augmentation to molecular-level assays, which excel at deep genotyping. The combination of both offers a powerful technique to render a global picture. Cell-level assays are often based on antibodies. Flow cytometry and time-lapse microscopy are two examples. Flow cytometry can perform cell sorting, to sort and count cells suspended in liquids according to their surface markers (i.e. cell membrane proteins). It is widely used in immune related topics, which need to analyze cells with or without particular surface markers such as CD4 and CD8. Time-lapse microscopy refers to the technique of stacking sequential images of multiple sampling points in time into a fast video (Beltman et al., 2009). It offers a valuable time component. Cells are usually stained with fluorescent tags targeting the protein of interest. The fluorescent signals in the image stack can be analyzed computationally.

5.2.1 Essential genes for cell cycles by time-lapsed microscopy

The control of the phase transition of cell cycles is performed by genes such as cyclins and cyclin-dependent kinases. However, they are not the only players. To find all cell cycle genes systematically, Neumann et al. (2010) employed a RNA interference method to knock down the known 21,000 protein coding genes one by one, and observe the cell behavior for two days using time-lapse microscopy. Human Hela cells were employed for this experiment. The histone 2B proteins were tagged with green fluorescent proteins to show the chromosomes. Computational image analysis was employed to classify various abnormalities of cell cycles. About 500 to 1000 genes were identified as the critical genes related to cell cycle functions. The data represent a valuable source of information, which has been made available on the Mitocheck website.

5.3 Conceptual frameworks from top-down

The top-down approach is the effective use of domain knowledge, organized in the form of conceptual frameworks, to direct research, to make abstractions and inferences, and also to guide data filtering and analysis. The importance of the conceptual framework escalates as the number of systems increases. Linnaeus's catalog of species is an example of a conceptual framework in biodiversity studies. In biomedical studies, conceptual frameworks on molecular mechanism and human physiology and pathology are particularly important to put fragmented knowledge into perspective, thereby revealing the essence of matter. Examples of conceptual frameworks include the Th1–Th2 balance for human immunology, the natural history of hepatitis B infection, and the hallmarks of cancer introduced below. Needless to say, the conceptual framework needs to be consistent with observation and empirical data. Furthermore, all conceptual frameworks have exceptions.

5.3.1 Hallmarks of cancer: a conceptual framework

The hallmarks of cancer are an excellent example of a conceptual framework. It was first conceptualized (Hanahan and Weinberg, 2000) by six major hallmarks, then further augmented into ten hallmarks based on accumulating evidence (Hanahan and Weinberg, 2011). The six major hallmarks are:

1. proliferation;
2. growth suppressor inhibition;
3. apoptosis evasion;
4. immortal replication;
5. angiogenesis; and
6. invasion and metastasis (Hanahan and Weinberg, 2000).

These hallmarks are largely caused by the endogenous molecular machinery of cells activated in the wrong place at the wrong time. These hallmarks also present in the progressive stages of the disease. The first hallmark is caused by oncogenic EGFR and Her2, which abnormally induce proliferative signals, then activate two downstream pathways,

Published by Woodhead Publishing Limited, 2013

one with the PI3K, Akt and PKB genes and the other with the RAS and MAPK genes. These oncogenes sustain a mitogenic, proliferative signaling in the cancer cells, accompanied by the malfunction of the endogenous inhibitory signals, such as PTEN, along the same route. The second hallmark is caused by malfunctions of endogenous tumor suppressor genes, such as RB and TP53. The third hallmark is caused by malfunctions of genes in the intrinsic and extrinsic apoptosis pathways. The fourth hallmark is closely related to attrition of the telomere (the two ends of chromosomes), which is caused by the fast proliferation of cancer cells. This can cause chromosome end-to-end junction, a mechanism for cancer to acquire more growth promoting mutations. As the disease progresses, telomerase becomes over-expressed, which elongates the length of the telomere, enabling it to continue replication infinitely. Two more hallmarks, genome instability and inflammation, were further proposed to be two underlying contributors of cancer (Hanahan and Weinberg, 2011). The reprogramming of energy metabolism and the evading of immune destruction were further considered as two emerging hallmarks (Hanahan and Weinberg, 2011).

The hallmarks of cancer offer invaluable guidance for future cancer-related research. Chapter 2 reported that a vast amount of somatic mutations has been found recently in tumor tissues. The mutations have complex patterns that vary among cancer types, disease stages, and even individual patients. This poses a serious challenge for distinguishing driver mutations (which play major roles in various stages of the disease) from passenger mutations. Furthermore, we may easily get lost in the large number of mutations and overlook important clues to useful knowledge. A conceptual framework can categorize the detected mutations into hallmarks, representing various stages of disease progression.

This conceptual framework was a result of prior top-down and bottom-up investigations. The individual evidence was accumulated empirically bottom-up and gradually connected into an integral. The global framework was gradually conceived top-down. The established conceptual framework offers both a functional scale and a timescale for the comprehension and treatment of this complex disease. The stages of cancer are now signified by distinctive hallmarks (Hanahan and Weinberg, 2011). The advanced stage of cancer usually manifests as the hallmark of metastasis, which often makes the disease intractable. Advanced cancerous cells can metastasize from one site to another, either through the bloodstream or through the lymph system. Cancer cells need to alter the surface property so as to enhance its invasion capability. Also, the metastatic cell needs to overcome an exotic, aquatic micro-environment with different osmotic pressure, when they enter the bloodstream. This could be a moment when these cells are vulnerable to attack if they are not yet prepared for such an environment. As a result, from a systematic point of view, metastatic cells are good therapeutic targets for late stage cancer.

5.4 Systems construction from bottom-up and top-down

5.4.1 SBMS data preprocessing: Integrators and filters

As was introduced briefly in Chapter 3, data filtering and integration are important analysis skills for modern omics data (Ideker et al., 2011). The integrators and filters are common signal processors in the engineering field. Here, it means the effective use of domain knowledge to filter the data of unnecessary complexity and to integrate data from heterogeneous sources.

5.4.2 Physiological and disease modeling

To fully characterize a system, two layers of information are required:

1. the system's static architecture in terms of the topology of networks;
2. the dynamical and collaborative behavior of the composition entities.

Mathematical equations are often used to describe quantitatively the internal and external characteristics of the systems. The closeness of predicted values to real values can be used to gauge whether a mathematical model adequately captures the essence of a biological system (Hartwell et al., 1999). Isolating the system from environmental effects, mathematically equivalent to setting all external variables to 0, is a common trick to focus only on the internal activities for easier comprehension. However, this approach may lose authenticity in the modeling of many biological systems, which are highly entangled with each other and manifest collective behaviors. Only a full appreciation of all layers can illustrate the complete picture of a complex system.

Published by Woodhead Publishing Limited, 2013

To model human physiology in health and disease and reveal important internal, unobserved variables is a major goal of SBMS. Such attempts are exemplified by the projects of Physiome and CancerSys, which model multiple organs such as the heart and liver.

It has been known that the sizes and numbers of first diagnosed primary and metastatic tumors inversely correlate with the patients' survival time. In theory, the value of the genomic mutation rate and of the metastasis rate determines the survival time. The two values cannot be directly measured, so they need to be estimated by observable parameters such as tumor sizes and growth rates. Considering a diseased person as a system, the genomic (metastatic) mutation rate and the metastasis rate are two internal variables of this system. A recent report by Haeno et al. (2012) showed the modeling of the cancer metastasis by a series of observable and internal parameters. The clinical data of 228 pancreatic cancer patients were used; 101 of them had autopsies, which offered critical numbers and sizes of metastatic sites. The models were then used to depict the survival distribution of an independent cohort (Figure 5.1).

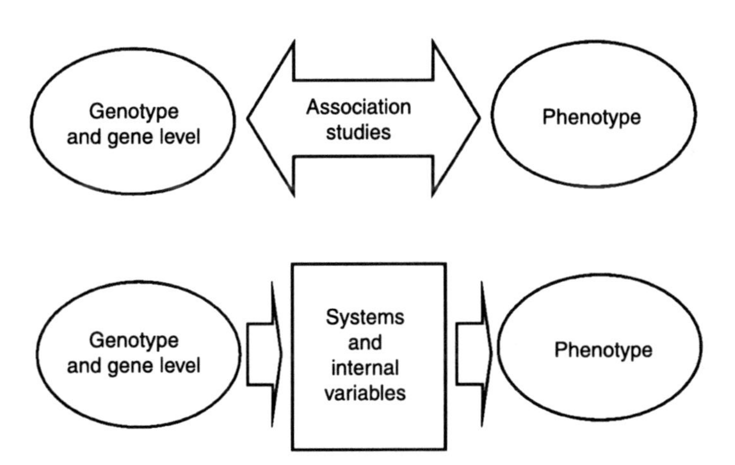

Figure 5.1 A schematic diagram of SBMS compared with association studies. Top: the association studies intend to find genes whose genotype or expression levels are associated with clinical phenotypes. Bottom: SBMS also intend to explain the clinical phenotypes, with observable genotypes and gene levels and internal variables estimated by systems modeling

5.4.3 Modeling and simulation of collective behavior

Apart from the modeling illustrated in the previous section, bottom-up modeling and simulation is useful for depicting the collective behavior of multiple systems in various spatial and temporal scales. A group of systems may manifest self-organizing capabilities. In an abstract illustration, a system is often depicted by a box with input and output signals (external variables) depicted by external arrows pointing toward and away from the boxes. Systems also have internal variables. The systems can also be modeled as depicted in the following pseudo code:

```
A system (inputs, outputs)
{
   internal variables: memory;
   outputs = function (memory, inputs);
};
```

5.4.4 Static characteristics of a system

In a confined system, the relationship between the composition entities can be represented as networks. Networks can be represented by two mathematically equivalent tools, the adjacent matrices and graphs (Barabási and Oltvai, 2004). An adjacent matrix is a square matrix where the rows and columns both represent the composition entities. The entries of the matrix could be binary, representing either connected (1) or unconnected (0). They could also be quantitative to represent the strength of connection, and the adjacent matrix is called the "weighted" matrix.

The topological properties of a network can be characterized by the various distributions of the node (the vertex) and the link (the edge), including:

Degree distribution

A degree (denoted as k) refers to the number of links associated with a node. The degree distribution, or histogram, is the counts of nodes with respect to various degrees $N(k)$. A degree distribution can also be shown

Published by Woodhead Publishing Limited, 2013

as *P(k)*, where the percentage of nodes is shown instead of the counts of nodes, simply by dividing all the counts with the total number of nodes in the network.

Average clustering coefficient

An average clustering coefficient *C(k)* indicates how tightly the nodes are linked to each other, for nodes with degrees *k*. For any given node, the clustering coefficient is the number of true links among its adjacent nodes, divided by the number of all possible pairwise links within a subset of nodes (Barabási and Oltvai, 2004).

The topological properties can characterize a biological network. The first typical network is a random network, where nodes are randomly joined by links.

The degree distribution *P(k)* of a random network usually follows a Poisson distribution. However, a scale-free network has the property that *P(k)* follows the power law approximately, where the logged *P(k)* linearly decreases as the logged degree *(k)* increases, manifesting as a straight line on a log-log plot. A hierarchical network is a special type of scale-free network commonly found in biological networks, where both P(k) and *C(k)* follows a power law.

5.4.5 Dynamic characteristics of a system

A living cell can switch between homeostasis states and transient states. Homeostasis states are essential for regular activities of the biological system. The transient states are often triggered by various extracellular stimuli such as endocrine hormones. A sequence of transient cell signaling and gene expression characterizes the transition states. It is thus essential to capture both the homeostatic and dynamic states of the cells in addition to the global topology of the network. The dynamic simulation known as flux analysis can be conducted in both the continuous and discrete domains with various degrees of granularity, most of which requires high-quality data with high time and space resolution for authentic modeling.

Differential equations are major tools of continuous domain simulation, where the concentration of particular receptors, ligands, enzymes or metabolites is modeled by the equations at various spatial and temporal scales. The use of simultaneous equations can capture the collective behavior of the composition entities. However, this approach usually results in too many equations to strenuously capture various interactions of composition entities behind the biomedical phenomena. Solving these equations is computationally expensive. One strategy to avoid the complexity of the whole system is to simulate the behavior of network motifs, such as various feedforward and feedback loops (Mangan and Alon, 2003; Burrill and Silver, 2010). It has been shown that the feedback motif can generate important biological properties such as the bistable (0 and 1) states (Farrell, 2008). Furthermore, a differential network strategy was proposed to focus the simulation of network behavior on the genes which show difference (obtained from contrast-level analysis) rather than simulating the entire set of entities (Peer and Hacohen, 2011).

One other approach is to employ higher level models when the collective behavior of systems is modeled. These systems (i.e. cells) are distinct entities but they share similar properties (i.e. a similar frequency of cell cycles). For example, the combination of Pott's model and Metropolis algorithm have been used to simulate cell sorting (Graner and Glazier, 2011), morphogenesis (Izaquirre et al., 2004), the behavior of malignant tumors (Pennacchietti et al., 2003) and the Tamoxifen treatment failure of cancer (Plank and Sleeman, 2003). The employment of Kuramoto models and their variants has been used to model the synchronization of cell cycles of unicellular phytoplankton organisms (Massie et al., 2010). This is a good example of emergence of collaborative behaviors from simple rules, particularly as synchronization is a prevailing characteristic of many biological systems.

Other choices of tools include discrete domain simulation such as petri nets and cellular automata, which can present the dynamic behavior with certain degrees of abstraction. Cellular automata are based on a spatio-temporal discrete lattice, therefore suitable for modeling of spatial information. The analysis of collective behavior of multiple self-organizing systems has been investigated under the framework of swarm intelligence.

Occasionally, two or more different types of cells are modeled. The attack of immune cells on tumor cells, and the counter-attack, require both cells to be modeled in order to illustrate the complex interaction.

Multi-scale simulations have been attempted to elucidate the multi-scale nature of biomedical phenomena, in such a way that a coarser scale simulates cell–cell interactions, while the other scale simulates molecular

Published by Woodhead Publishing Limited, 2013

activities within cells. Abstract data structures, such as the quadtree structure of multiple layers, are 2-D spatial grids. Activities on the two scales may take place at different paces (timescales).

Recent evidence suggests that even in meticulously controlled experimental conditions in a laboratory, a degree of randomness and variation (or noise) was still observed (Raj et al., 2010). Noise has been perceived as one driving force of a biological phenomenon (Eldar and Elowitz, 2010), therefore, in addition to the simulation using deterministic equations and modeling techniques, we may add a stochastic component to authentically present a biological fact.

5.4.6 Cell simulation frameworks

The E cell and virtual cell focus on the molecular and biochemical level within cells, addressing the dynamics of signal transduction, regulatory and metabolic networks. The sub-cell compartmental model is constructed and integrated gradually, so as to simulate a particular facet of cells. The Epitheliome project is an example of tissue-level simulation, aiming to depict the epithelial cell growth and the social behavior of cells in culture. The different scales of simulation can shed light on different aspects of life.

5.5 Specific directions of systems biomedical science

5.5.1 Human microbiome, metagenomics and viral host interactions

The first example of SBMS is to investigate the human genome and microbiome together to understand their interactions on human health status. Our body is intimately influenced by external and internal environments with many residential microbial organisms. Counting virus, bacteria, archaea and fungi all together, approximately 100 trillion microorganisms live inside the human body with a total of 100-fold more than humans (Ley et al., 2006; Qin et al., 2010; Arumugam et al., 2011). They are commensal, neutral or even pathogenic to human health, and the effects are mostly elusive. Evidence shows that inflammatory diseases, such as Crohn's disease and ulcerative colitis, are caused by pathogenic host and microbial interactions (Virgin and Todd, 2011). Crohn's disease is a consequence of concurrent risk factors on host gene variants, viral infections, toxins and commensal bacteria (Cadwell et al., 2010). It is thus very important to put microbial organisms into perspective, with about the same weight as the human genome.

Metagenomics is the concurrent analysis of multiple genomes. One typical example is the use of NGS for the study of multiple organisms in a specific field environment such as sea water and ground soil. This approach would also be very useful for biomedical study on the interactions of pathogens, residential microbials and the human body. This approach has already rendered insights into, for example, the gut microbiome by ways of analyzing human fecal samples (Arumugam et al., 2011; Qin et al., 2010). In contrast to traditional capillary sequencing, where cell isolation and culture are important steps in the protocol, the metagenomic approach directly collects samples, conducts shotgun sequencing of all organisms simultaneously, and then analyzes the gene contents (and abundance) as a whole. Often the first goal is to get a snapshot of the entire gene pool, followed by the assembly of individual genomes as the second goal. Tools and database resources are in great demand in this fast moving field.

Infectious diseases remain major challenges to human health. Genomic information of pathogens such as the virus or bacteria are valuable for the understanding of their interaction with the host. Assays on the pathogen genome can help us to identify the sub-strain of pathogens. Many viral and bacterial mutations are also shown to be responsible for drug resistance. This genomic information could contribute to better management of pathogen induced diseases. For example, many mutations in the pre-S or pre-core regions of the hepatitis B viral genome have been shown to be positively associated to the occurrence of host liver fibrosis/cirrhosis and even liver cancer. Mutations in the S protein may cause resistance to the antiviral drug Lamivudine. Hence, the detection of these mutations could represent a change of medication for better therapeutic effects.

5.5.2 Molecular model of embryo development

The development of an embryo (i.e. ontogeny) and the evolution of species (i.e. phylogeny) are two separate fields of study, which share

Published by Woodhead Publishing Limited, 2013

ambiguous yet intricate connections. It has long been observed that during a special period of animal embryonic development, called the "phylotypic" stage, the morphologies and anatomies of distinct species appear very similar (Prudhomme and Gompel, 2010; Domazet-Loso and Tautz, 2010; Kalinka et al., 2010). This morphological resemblance (and divergence) among species during embryonic development has been described by an hourglass analogy: the high resemblance (low variety) in the phylotypic stage is shown as the middle, narrow part of an hourglass; the high divergence in the beginning and late stage of embryonic development is shown as the wide parts. Such an analogy has remained a subjective, phenotypic description for years.

Molecular evidence for the hourglass model of development has only been established recently (Prudhomme and Gompel, 2010; Domazet-Loso and Tautz, 2010; Kalinka et al., 2010). Domazet-Loso and Tautz (2010) conducted a joint genome-wide DNA, RNA and protein level analysis across species, fitting into our definition of systems level analysis. It also featured a development scale time course analysis. First, transcriptomics profiles were measured during different stages of zebrafish development, including embryo, larva, juvenile to adult stages. This was done by quantitative measurements of all transcribed RNAs using Agilent microarrays every two hours, a very fine-grained series of time points.

Second, a phylogeny of 14 evolutionary stages was constructed, starting from the unicellular organism to the zebrafish. Each stage was represented by full genomic (DNA) sequences of one or more species supplemented by the RNA level expressed sequence tag (EST) data. The main purpose of this phylogeny is to provide a time index corresponding to the stages, where 1 represents the first stage of the origin of cell, and 14 represents the last stage of the zebrafish.

Third, the time index of each gene of the zebrafish was traced back, based on protein sequence homology to earlier species in phylogeny. The revised time index represents the first emergence of a gene in phylogeny. The smaller the index, the earlier the emergence of this gene in time.

Fourth, the authors defined a phylogenetic time index for each embryo developmental stage, called the transcriptome age index (TAI). It is basically a weighted average of time index of all genes by their expression levels. When plotting the TAI with respect to the developmental stages, the curve actually resembles the hourglass model, with the valley of the curve occurring at the embryonic segmentation stage (~23 hours after fertilization). This shows that the phylotypic stage expresses the oldest gene set, conserved across species, to possibly dictate the body plan of development.

It is interesting that the authors stopped here after the joint analysis of DNA, RNA and protein information. One way to continue this project is to conduct the module level annotation of the genes in the phylotypic stage by enrichment analysis (Section 3.4 in Chapter 3). In this way, they could further confirm whether the genes at this stage actually play pivotal roles for body plans of development, inspired by the success of the molecular hourglass model.

5.5.3 Diseases grouping, subtyping and interactions

Optimum medical intervention relies on a systematic approach, featuring a balanced consideration of all human physiology, pathology and the molecular basis of disease. This is a basic requirement a physician needs to make a treatment decision based on the current best knowledge. Nevertheless, systems level consideration still has room for improvement, primarily due to insufficient knowledge. For example, several diabetic drugs on the market are designed to stimulate the patients' pancreas to secrete more insulin, thereby increasing the absorption and utilization of blood sugar by muscle and liver cells. However, over-stimulation of the pancreas may result in an elevated risk of pancreatic cancer in the long term (Currie et al., 2009). For Type 2 diabetes patients, whose main problem is the reduced sensitivity of muscle and liver cells to insulin (i.e. insulin resistance), then insulin stimulating drugs may not be adequate. Rather, insulin sensitizers may be a better class of drugs for these particular patients by targeting the real problem.

Conventionally, a disease is often investigated as an isolated system with its distinct diagnosis and treatment strategies. Meanwhile, it has long been observed that many diseases have shared etiology and risk factors, which may reflect their tangled underlying mechanisms. Epidemiological data shows that obesity elevates the risk of diabetes, which in turn elevates the risk of vascular complications such as nephropathy, retinopathy, diabetic foot, and most worrisome, coronary artery disease or heart attack. These diseases, together with other clustered symptoms such as hypertension, has been discussed under the general conceptual framework of metabolic syndrome (Reaven, 1988, 1993, 2005; Taniguchi et al., 2006). Insulin resistance is hypothesized as a shared etiology of phenotypes ascribed to metabolic syndrome, providing insight into the whole picture. Adding to this complexity is the

Published by Woodhead Publishing Limited, 2013

fact that there are diabetes patients who are not obese and there are cardiovascular patients who have no trace of diabetes. Thus, a subtyping of conventional disease definition may also be required.

Box 5.1 Useful bioinformatics tools and resources for systems biomedical science

A. Data Visualization with graphs

- Cytoscape

Cytoscape is a free software platform dedicated to network analysis and visualization. The basis of the platform offers a convenient way to construct networks and graphs. On top of the base, many plug-in applications have been constructed, offering a spectrum of toolkits such as BiNGO and PiNGO for graphical gene ontology annotation.

- Grfaphviz

Grfaphviz is an open source graphical presentation software, which can be used to illustrate a wide variety of networks.

- Cell Collective

Cell Collective is a simulation software for the dynamics of multiple genes.

The prevention of a disease and the medical care of patients also require the whole picture of disease subtypes and their interactions. Disease grouping and subtyping can be sustained by molecular evidences. For example, mutations in the low density lipoprotein receptor (LDLR) gene may elevate the risks of both the familial hypercholesterolemia and coronary atherosclerotic heart disease, the former of which is thought to be a rare, familial disease while the latter the common complex disease (Lupski et al., 2011). There is new thinking on globally pictured "disease systems", which either divides a conventional disease into many subtypes, or merges many diseases into a larger system (Virgin and Todd, 2011).

5.6 Case studies

5.6.1 *The omics of one person predicts the onset of T2D*

Chen et al. (2012) have explored the idea of prospectively collecting extensive time course, various omics data from a single person, Snyder, who is also the senior author of the published work. A spectrum of genomics, transcriptomics, proteomics and metabolomics was collected together with basic physiological parameters (i.e. weight) and biochemistry measurements (i.e. Hba1C).

5.6.2 *The phase-locking model of cell cycles*

Cell cycles are endogenous, rhythmic processes of living cells. By going through a cycle, a cell divides into two daughter cells. A cell cycle comprises distinct phases, G1, S, G2 and M, where the S phase represents the process of chromosomal duplication and the M phase represents cell division. G1 and G2 are intermediate steps toward S and M, respectively. The progression of the cell cycle has long been depicted as one large mechanistic ratchet model, where a rigid control and checkpoints of the progression of phases are in place, because the next phase cannot be activated until the previous phase is fully completed. However, this perception has not been fully validated.

Lu and Cross (2010) proposed an alternative model where the cell cycle is a synchronization of multiple molecular cycles, which are gradually emerging independently in evolution history. Synchronization is achieved by the phase-locking of these cycles, manifesting a coherent state-switching behavior. To demonstrate this concept, Lu and Cross designed a series of experiments and de-coupled the molecular links between different molecular cycles in yeast cells. Each molecular cycle then demonstrated a unique cycling frequency and became out of phase with the main cycle.

The phase-locking model offers a conceptual level insight on how the system of a cell operates mechanically. In the future, we may be able to classify genes into modules according to molecular cycles, and identify genes as “hinges” based on their molecular roles in the synchronization of two molecular cycles.

Published by Woodhead Publishing Limited, 2013

5.7 Take home messages

- Advancement of biomedical science requires thinking outside the box and crossing conventional boundaries.
- Internal variables of a modeling system can be used to indicate internal states, thereby inferring a dynamic phenotype.
- Conceptual frameworks such as the hallmarks of cancer can offer a top-down panoramic view of the problems and also guide data analysis.
- Adaptive models, such as agent based modeling or swarm intelligence, can be used to simulate emerging effects from bottom-up.
- Filters and integrators are useful for distilling valuable knowledge from a deluge of data.

5.8 References

Anderson, P.W. (1972) More is different. *Science, New Series*, **177(4047)**: 393–6.

Arumugam, M., Raes, J., Pelletier, E., et al. (2011) Enterotypes of the human gut microbiome. *Nature*, **473(7346)**: 174–80.

Barabási, A.L. and Oltvai, Z.N. (2004) Network biology: understanding the cell's functional organization. *Nat. Rev. Genet.*, **5(2)**: 101–13.

Beltman, J.B., Marée, A.F. and de Boer, R.J. (2009) Analysing immune cell migration. *Nat. Rev. Immunol.*, **9(11)**: 789–98.

Burrill, D.R. and Silver, P.A. (2010) Making cellular memories. *Cell*, **140**: 13–18.

Cadwell, K., et al. (2010) Virus-plus-susceptibility gene interaction determines Crohn's disease gene Atg16L1 phenotypes in intestine. *Cell*, **141(7)**: 1135–45.

Chen, R., Mias, G.I., Li-Pook-Than, J., et al. (2012) Personal omics profiling reveals dynamic molecular and medical phenotypes. *Cell*, **148(6)**: 1293–307.

Currie, C.J., Poole, C.D. and Gale, E.A. (2009) The influence of glucose-lowering therapies on cancer risk in Type 2 diabetes. *Diabetologia*, **52(9)**: 1766–77.

Domazet-Lošo, T. and Tautz, D. (2010) A phylogenetically based transcriptome age index mirrors ontogenetic divergence patterns. *Nature*, **468(7325)**: 815–8.

Eldar, A. and Elowitz, M.B. (2010) Functional roles for noise in genetic circuits. *Nature*, **467(7312)**: 167–73.

Ferrell, J.E. Jr. (2008) Feedback regulation of opposing enzymes generates robust, all-or-none bistable responses. *Curr. Biol.*, **18(6)**: R244–5.

Graner, F. and Glazier, J.A. (2011) Simulation of biological cell sorting using a two-dimensional extended Potts model. *Phys. Rev. Lett.*, **86**: 4492–5.

Haeno, H., Gonen, M., Davis, M.B., Herman, J.M., Iacobuzio-Donahue, C.A. and Michor, F. (2012) Computational modeling of pancreatic cancer reveals kinetics of metastasis suggesting optimum treatment strategies. *Cell*, **148 (1–2)**: 362–75.

Hanahan, D. and Weinberg, R.A. (2000) The hallmarks of cancer. *Cell*, **100**: 57–70.

Hanahan, D. and Weinberg, R.A. (2011) Hallmarks of cancer: The next generation. *Cell*, **144**: 646–74.

Hartwell, L.H., Hopfield, J.J., Leibler, S. and Murray, A.W. (1999) From molecular to modular cell biology. *Nature*, **402**: C47–C52.

Ideker, T., Dutkowski, J. and Leroy Hood, L. (2011) Boosting signal-to-noise in complex biology: Prior knowledge is power. *Cell*, **144**: 860–3.

Izaquirre, J.A., Chaturvedi, R., Huang, C., et al. (2004) CompuCell, a multi-model framework for simulation of morphogenesis. *Bioinformatics*, 20(7): 1129–37.

Kalinka, A.T., Varga, K.M., Gerrard, D.T., Preibisch, S., Corcoran, D.L. et al. (2010) Gene expression divergence recapitulates the developmental hourglass model. *Nature*, **468(7325)**: 811–4.

Ley, R.E., Peterson, D.A. and Gordon, J.I. (2006) Ecological and evolutionary forces shaping microbial diversity in the human intestine. *Cell*, **124(4)**: 837–48.

Lu, Y. and Cross, F.R. (2010) Periodic cyclin-Cdk activity entrains an autonomous Cdc14 release oscillator. *Cell*, **141(2)**: 268–79.

Lupski, J.R., Belmont, J.W., Boerwinkle, E. and Gibbs, R.A. (2011) Clan genomics and the complex architecture of human disease. *Cell*, **147(1)**: 32–43.

Mangan, S. and Alon, U. (2003) Structure and function of the feed-forward loop network motif. *Proc. Natl. Acad. Sci. USA*, **100(21)**: 11980–5.

Massie, T.M., Blasius, B., Weithoff, G., Gaedke, U. and Fussmann, G.F. (2010) Cycles, phase synchronization, and entrainment in single-species phytoplankton populations. *Proc. Natl. Acad. Sci. USA*, **107(9)**: 4236–41.

Neumann, B., Walter, T., Hériché, J.K., et al. (2010) Phenotypic profiling of the human genome by time-lapse microscopy reveals cell division genes. *Nature*, **464(7289)**: 721–7.

Peer, D. and Hacohen, N. (2011) Principles and strategies for developing network models in cancer. *Cell*, **144**: 864–73.

Pennacchietti, S., et al. (2003) Hypoxia promotes invasive growth by transcriptional activation of the met protooncogene. *Cancer Cell*, **3**: 347–61.

Plank, M.J. and Sleeman, B.D. (2003) A reinforced random walk model of tumour angiogenesis and anti-angiogenic strategies. *Math. Med. Biol.*, **20(2)**: 135–81.

Prudhomme, B. and Gompel, N. (2010) Evolutionary biology: Genomic hourglass. *Nature*, **468(7325)**: 768–9.

Qin, J., Li, R., Raes, J., et al. (2010) A human gut microbial gene catalogue established by metagenomic sequencing. *Nature*, **464(7285)**: 59–65.

Raj, A., Rifkin, S.A., Andersen, E., et al. (2010) Variability in gene expression underlies incomplete penetrance. *Nature*, **463(7283)**: 913–8.

Reaven, G.M. (1988) Banting Lecture 1988. Role of insulin resistance in human disease. *Diabetes*, **37(12)**: 1595–607.

Reaven, G.M. (1993) Role of insulin resistance in human disease (syndrome X): An expanded definition. *Annu. Rev. Med.*, **44**: 121–31.

Published by Woodhead Publishing Limited, 2013

Reaven, G.M. (2005) The metabolic syndrome: requiescat in pace. *Clin. Chem.*, **51(6)**: 931–8.
Taniguchi, C.M., Emanuelli, B. and Kahn, C.R. (2006) Critical nodes in signalling pathways: insights into insulin action. *Nat. Rev. Mol. Cel. Biol.*, **7**: 85–96.
Virgin, H.W. and Todd, J.A. (2011) Metagenomics and personalized medicine. *Cell*, **147(1)**: 44–56.

6

Clinical developments

DOI: 10.1533/9781908818232.129

Abstract: Modern medicine has progressed alongside the accumulation and refinement of knowledge of human biology and pathology. This book has shown how new knowledge can be obtained by studies of deep phenotyping and various omics approaches. The next critical challenge is to efficiently translate this knowledge into clinical practices, including diagnostics, prediction methods and therapies. We will explore the clinical developments from three angles:

1. how to define unmet medical needs and also fulfill them;
2. how to develop clinical products;
3. how clinicians can employ the current knowledge to facilitate their practice.

Three different types of genetic tests were used as a demonstration. These tests were unavailable until a decade ago.

Key words: translational medicine; status quo, prediction, clinical utility, sensitivity, specificity, positive predictive value, negative predictive value.

6.1 Fulfilling unmet medical needs

Despite a large increase in the volume of medical knowledge in modern times, many unmet medical needs remain, which can be broadly categorized into three types:

1. better diagnostics;
2. better predictions of future clinical events and outcomes;
3. better treatments.

The importance of diagnosis for clinical practice cannot be over-emphasized. Only when diseases are precisely diagnosed, can they be properly treated. As there are hundreds or even thousands of commonly encountered diseases with a wide spectrum of personal variation in severity, it is very challenging for a clinician to accurately diagnose the disease in a short time, and come up with a personally optimized treatment strategy. Furthermore, many diseases have very subtle, easily overlooked symptoms, meaning they are diagnosed when they are already at a late stage. An improvement would be if we could devise new diagnostics tools, which are more clinically accessible and sensitive at the earlier stages.

Apart from diagnostics, precise prediction of future clinical outcomes, such as treatment responses, disease progression or relapses, is no less important. A prediction of different courses of disease progression under various treatments would be extremely useful for optimizing personal treatment strategies.

To fulfill these unmet needs, we first need to define them. Thus, an efficient use of new knowledge is required to advance clinical practice with earlier diagnosis and smarter medical intervention.

6.1.1 Status quo vs. future events

In previous chapters, we saw that the combination of multiple omics technology can illuminate the molecular workings of biological systems. They are equally powerful in addressing specific unmet medical needs. But what type of study is adequate, and which is the right omics platform? We must first be clear about whether the need is about the "status quo", such as the current disease state or the underlying mechanisms, or about the future events, such as disease onset or progression. A typical transcriptomics or proteomics study involves the examination of biological specimens under different conditions, for example, cancer vs. non-cancer. The detected gene expression differences may reflect the underlying disease mechanism or indicate the current stage (or severity) of the disease. This so-called "status quo" investigation can be used to reveal currently altered genes and decipher disease mechanisms. New treatment strategies and diagnostic methods can then

Published by Woodhead Publishing Limited, 2013

be derived. However, the disease vs. non-disease comparison is often not useful for capturing biomarkers to predict disease events ahead of time, because the detected gene alteration often occurs after rather than before the onset of the disease (Figure 6.1). To capture predictive biomarkers, a prospective study is often required to correlate the baseline gene (i.e. biomarker) alterations with imminent clinical events occurring later on. Another approach is by the use of germline variants. Genotypes of germline variants are fixed at birth, presumably ahead of disease onset.

6.2 Translational medicine

The utilization of existing biomedical knowledge into medical practice is called "translational medicine". One typical scenario is to validate the promising hypotheses derived from exploratory cell-line experiments, animal models or clinical observations, using human samples with adequate sample size, which is no trivial task. Multiple steps are required in sequence for translational medicine. First, if the prior evidence is established on cell lines or animal models, corresponding human studies need to be conducted. Alternatively, if the exploratory knowledge is already derived from retrospective clinical observations or from prospective clinical studies, then we can move on to the next step (validation) directly.

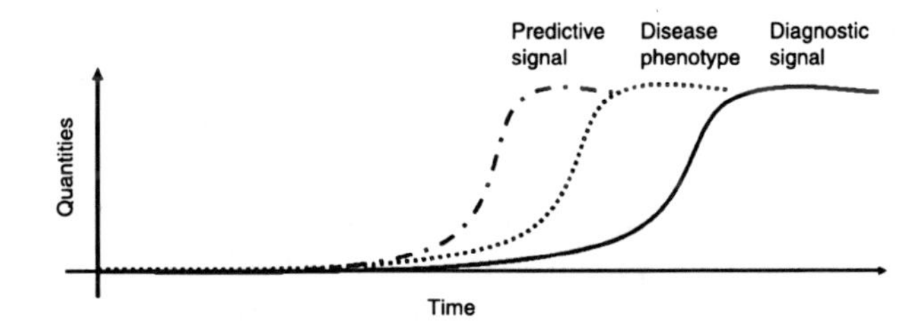

Figure 6.1 A time-course illustration of disease phenotype and biomarker quantities. The dotted curve shows the onset and progression of the disease phenotype. The solid curve represents a quantitative biomarker, which is often detected in comparison of diseased and non-diseased samples. This biomarker may offer diagnostic signals. The dash-dot curve represents a prediction biomarker which arises ahead of disease onset. This type of biomarker usually cannot be found in conventional case vs. control studies

Second, a validation of the exploratory findings is required. This could be done in either the retrospective study of an independent cohort of patients, or prospective study, depending on the task. Prospective validation is usually required to establish the drug efficacy biomarkers (Patterson et al., 2011).

Third, before moving on to clinical trials, the practical format of the product, such as the drug, treatment, biomarker device, etc. needs to be designed and manufactured (and even approved by authoritative agencies) with consistent performance. Finally, a series of adequately powered clinical trials are needed to test the product in human subjects so as to demonstrate its clinical use.

6.2.1 Challenges and tips for translational medicine

The first challenges encountered in translational medicine are about effect size. Exploratory findings are often justified by small P-values, representing that observed biological effects are genuine as opposed to random fluctuations. P-values are defined statistically as the probability of rejecting the null hypothesis when the null hypothesis is true. In other words, P-values are related to the probability of false positives (Fp). They do not quantify the effect size. A genuine etiology of disease will be seen as a small P-value, as long as the sample size is sufficient, no matter how small the effect of the size is. This will pose a problem for clinical utilization because we need an effect which is both "genuine" and "large" for practical clinical applications. Since most claimed biomarkers capture genuine but small effects, a collection of them are often required to cover most of the important etiologies of a certain subtype of disease. Even though such a collection of biomarkers (also called biosignatures) can be identified, we still need to combine them and construct clinically useful prediction models and demonstrate their performance. Multivariate analysis is thus required and performance indexes need to be established.

The second challenge is that the validation may require prospective studies, as opposed to the retrospective study design which is often employed in exploratory studies. A prospective study means that the classification or prediction is made (often at baseline) before the clinical result is manifested. A review of FDA approved biomarkers prediction treatment effects shows that efficacy related biomarkers are mostly validated by prospective studies, while the safety related biomarkers are validated by retrospective studies (Patterson et al., 2011). Survival

Published by Woodhead Publishing Limited, 2013

analysis is particularly useful for analyzing longitudinal data often produced by prospective studies.

The third caveat on translational medicine is the need to replace relative values with absolute values. Gene expression differences obtained by exploratory transcriptomic and proteomic studies are often based on relative values. We need to estimate the absolute gene abundance by different quantitative assays. The conversion of assay platforms means that some original observed effects may not be seen in new platforms.

The fourth caveat is that translational medicine may only improve scientific knowledge but cannot guarantee clinical application in a short period of time. For example, mutations on CFTR were discovered to be responsible for cystic fibrosis in 1989 (Rommens et al., 1989). Since then, a variety of translational medicine strategies has been imposed but none has actually resulted in a routine therapy so far, despite a drastic increase of knowledge on ion channels (Pearson, 2009). Hopefully, the knowledge can be successfully translated into clinical use in the near future.

6.2.2 Regression based methods for biosignature

Many biomedical investigations are designed to detect clinical variables (e.g. age, gender, body weight), biochemistry measurements (e.g. serum CRP levels), genomic variants, gene expression levels, or peptides abundance, which show significant differences in different clinical conditions (e.g. health and disease). A step toward clinical use is to classify subjects into clinical conditions based on the variables. Often one variable is not enough to classify subjects correctly, because of limited effect size. In biomedical terms, this variable captures only one of the etiologies of the complex phenotype. A combination of multiple parameters may achieve a better result of classification. Regression based methods are one class of multivariate analysis, which has two general purposes:

1. to assess the joint effect;
2. to identify confounding effects among variables.

These methods include linear regression, logistic regression, ordinal regression and Cox regression. For example, a logistic regression equation can depict the relationship between clinical phenotypes (dependent variable) and abundance of peptides (independent variables).

How many independent variables (e.g. number of serum proteins) are suitable for a logistic regression model? This actually depends on the sample size of the two groups. An empirical index, called events per variable (EPV) is defined as the sample size of the smaller group divided by the number of variables. Peduzzi et al. (1996) suggested that EPV should be larger than 10.

The independent variables can be both quantitative and binary. Hence, they have the potential to combine clinical variables, proteomics, transcriptomics and genotype data.

6.2.3 Survival analysis

So far we have seen analysis of association between omics variables and clinical phenotypes. Such a study basically examines the cross-sectional connection, where the number of subjects in clinically distinct groups provides the power for statistical analysis. Though useful, this type of analysis does not use a very important dimension of the patient: the time dimension. In fact, many biomedical effects are illustrated longitudinally where time is an important factor. Examples include the expected time of tumor relapse after surgery, the expected length of time before death, etc.

Survival analysis is the analysis of such data. It has three main uses:

1. presentation and visualization of longitudinal data;
2. the prospective (and also retrospective) demonstration of the difference in clinical progression of distinct groups;
3. the exploration of associated markers.

A survival analysis starts with the definition of events, which are critical phenotypic change and the time to which this is evaluated. An event could be, for example, tumor relapse, death, an enlargement of tumor, etc. Evaluation of overall survival is by the event of death. Overall survival is particularly important in clinical studies of cancer, as the extension of survival time is a critical goal for late-stage cancers.

Kaplan–Meier curves are important tools for the presentation and visualization of longitudinal data in survival studies (Figure 6.2). The horizontal axis shows the time (months) to event. The vertical axis shows the percentage of subjects who have not yet encountered the event.

An incidence of event is revealed as a sudden drop in the curve. A Kaplam–Meier plot can show a single curve or multiple curves, representing different subject groups, such as patient groups stratified by a biomarker. Survival analysis could be used for both the exploration and validation of potential clinical biomarkers. Statistical tests, such as the

Published by Woodhead Publishing Limited, 2013

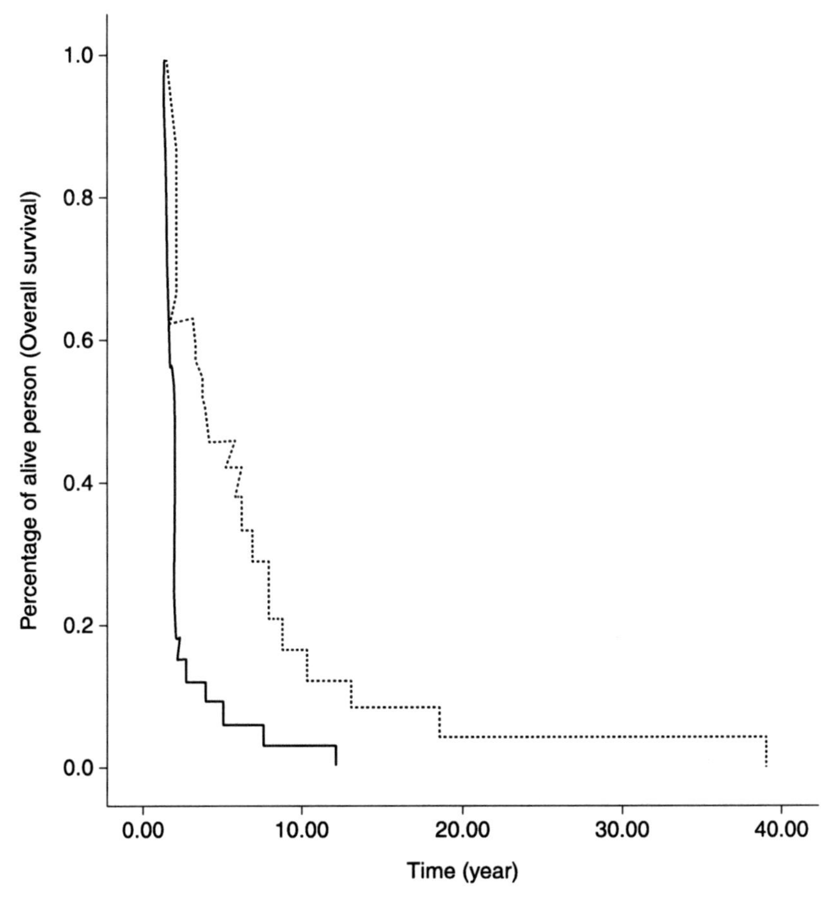

Figure 6.2 A typical Kaplan–Meier plot of survival curves. Solid and dotted curves represent two different groups of biomarker stratified patients

log-rank test and the Wilcoxon test, could be used for comparison of two groups. The tests examine whether two curves are different, representing different clinical manifestation in time of two groups.

Exploration of associated markers is basically done by the Cox proportional hazard model when the variables are continuous (as the gene expression measurements). Cox regression can also be used when there are two or more independent variables.

In survival analysis, two terms are frequently confused, the "progression free survival" and "time to progression". They are different when death occurs before progression. The former considers it as an event, while the latter considers it as censored data (Saad and Katz, 2009).

6.3 Clinical product development

6.3.1 *Performance indexes*

Clinical products such as drugs, prediction assays and diagnostic devices all need to demonstrate clinical availability and clinical utility before they can be used. A diagnostic device also needs to present assay utility. Clinical performance indexes reflect the level of clinical utility. The indexes include sensitivity, specificity, positive predictive value (PPV) and negative predictive value (NPV). They are based on the counts of samples, which were accurately or erroneously classified by a diagnostic or prediction test. Four important counts include Tp (true positives), Tn (true negatives), Fp (false positives) and Fn (false negatives). The subjects with positive clinical status are either truly declared as positives (Tp) or falsely declared as negatives (Fn). However, subjects with negative clinical status are either truly declared as negatives (Tn) or falsely declared as positives (Fp). The total number of predicted positives is Tp+Fp. The total number of predicted negatives is Tn+Fn (Table 6.1). The total number N = Tp + Tn + Fp + Fn.

The sensitivity is defined as the portion of Tp within all clinical positives, i.e.:

$$\text{Sensitivity} = Tp/(Tp + Fn)$$

Similarly,

$$\text{specificity} = Tn/(Tn + Fp)$$

PPV is defined as the portion of Tp within all the predicted positives, i.e.:

$$PPV = Tp/(Tp + Fp)$$

Similarly, NPV is defined as

$$NPV = Tn/(Tn + Fn).$$

Table 6.1 **The 2 × 2 contingency table of classifications**

	Clinical Status		
	Negative	**Positive**	**Sum**
Negative	Tn	Fn	Tn+Fn
Prediction Positive	Fp	Tp	Tp+Fp
Sum	Tn+Fp	Fn+Tp	n

Published by Woodhead Publishing Limited, 2013

The importance of indexes and their acceptable ranges depends on clinical applications. Sensitivity may be more important than specificity if the assay is developed for an early diagnosis of cancer, and the positive results can be confirmed by further medical examinations. A typical example is the prediction of the adverse drug reaction known as the Steven Johnson syndrome by an HLA allele. The sensitivity, specificity and NPV have all reached above 90%, an extraordinary performance. Yet the PPV is very poor, at less than 10%. This is because the particular HLA allele is rare in the population. This does not hurt the clinical application of this test, because as long as the test is negative, we are confident that the adverse drug reaction will not occur due to the high NPV of this test.

It is important to note that the proportion of positive cases can affect the values of PPV and NPV. In such cases, meaningful PPV and NPV are derived only when the proportions of positive cases are similar to the incidence, defined as the risk of developing a clinical status within a specified period of time. A prospective study is usually preferable for estimating PPVs and NPVs, because the clinical status automatically follows the incidence.

A diagnostic test usually gives a quantitative measurement, and a cutoff value is used to decide which range will be considered as positive and which range as negative. For example, when the logistic regression equation is used to render a probability of (positive) occurrence, which is in the range 0 and 1, then a cut off in between can dissect the range into positive and negative zones. The performance value is thus a trade off between these indexes, which is controlled by a cut off value. At one extreme, if we assign all subjects to be positive by setting the cut off at 0, then sensitivity will be 100%, yet specificity will be 0%. Alternatively, if the cut off is to be 1, then sensitivity will be 0%, yet specificity will be 100%.

Using a series of cut offs (usually with a constant increment such as 0.1), a series of corresponding performance indexes can be generated and plotted in a receiver operating characteristic (ROC) curve, an 2-D plot where the vertical axis is the sensitivity and the horizontal axis is 1-specificity. The area under the curve (AUC) is a value between 0 and 1, which estimates the general classification performance using a series of cut offs.

In practice, a test still requires a fixed value of cut off so as to be clinically useful. The cut off can be determined by maximizing one of the following combined measurements:

1. *Accuracy*: the proportion of accurately predicted subjects amongst all subjects, i.e. (Tp + Tn)/N.

2. *The sum of sensitivity and specificity*: the goal of which is to maximize both indexes.
3. *Youden index*: which is sensitivity + specificity −1.
4. *F1 score*: Tp/(Tp + Fp + Fn) (note that there is no Tn in the equation).
5. *Matthew correlation coefficient*: (Tp*Tn–Fp*Fn)/sqrt((Tp + Fn) *(Tp + Fp)*(Tn + fp)*(Tn + fn))

6.3.2 Clinical trials

Clinical trials are critical stages that demonstrate comparatively better performance and manageable adverse reactions of clinical products. They are categorized into four consecutive phases. A phase-one trial usually involves healthy individuals to test the potential toxicity of investigational new drugs (INDs) on the human body. A phase-two trial usually includes a dose escalating study to examine the best dose in terms of maximized drug effects and minimized adverse reactions. A phase-three trial aims to demonstrate the superior or non-inferior treatment effect of the IND compared with a placebo or existing drugs. This is done by comparison of different treatment protocols called arms. As the treatment of the placebo may have ethical issues, many current trials choose to compare the drug with existing drugs and try to establish superior or non-inferior effects and with less adverse drug reactions. A phase-four trial is often conducted after the drug is already on the market.

Clinical trials are designed to ensure objectivity. Basically, trial subjects need to fulfill several defined criteria (called inclusion criteria) before they can be enrolled. Then they are randomly assigned to trial arms in randomized clinical trials so as to guarantee objectivity. Double-blind studies represent another measure to ensure objectivity if possible. Both the doctor and the patients do not know which arm the patient is assigned to. Also, the pharmacist who prepares the drug does not know which specific patient the drug is to be given to. The clinical design needs to consider the practical applicability of the design. If a clinical trial aims to compare drugs of different types of administration, such as an oral drug and a subcutaneous injection drug, then the patients and doctors will know which drug the patient is taking.

In terms of statistical analysis there are two different types, the intended to treat analysis and the per protocol analysis. The major difference of the two types is on how to deal with subjects who have dropped from the analysis or do not quite follow the protocol. The former will include all

Published by Woodhead Publishing Limited, 2013

subjects who are intended to be treated according to the protocol into the final analysis, while the latter will remove those subjects who do not follow the protocol for various practical reasons.

It is common practice to estimate the sample size of each arm before the trial. The estimation will be based on information such as effect size obtained from previous studies.

6.3.3 Biomarker and drug co-development

This is a lengthy process of drug development from cell-based assays, animal models to the clinical trials, which usually takes more than a decade. Unfortunately, many INDs are failed at various stages of development, and the fail rates are high. Cancer is the field with the highest fail rate since IND (>85%). One reason for such a high fail rate is that the disease is often heterogeneous and the drug is only effective to a subtype of the heterogeneous disease. One possible solution is to employ biomarkers, which can screen the right subgroup of patients for the drug. It is expected that such a design can more easily demonstrate the effect of a drug, thereby improving its success rate.

In the past decade, we have witnessed a rapid progress in Pharmacogenomics, the study of genetic effects on the efficacy and toxicity of treatment toward various diseases. The progress is particularly prominent in the findings of associations of germline variants on drug metabolizing enzymes (e.g. the Cytochrome P450 family) to the half life of small molecule drugs, and in cancer treatments where the somatic mutations on the drug targets (e.g. EGFR, HER2) and their down streams (kRas) in the cancer tissue play a large role in drug responses. The search of genomic variants and their relationship with protein–drug responsiveness are exemplified by the responsiveness of interferon treatment to Hepatitis B patients.

6.4 Critical use of clinical information

Clinical reports are literally produced daily. New concepts are constantly replacing old concepts, which have been accepted for a long time but then turned out to be wrong. A clinical practitioner needs to employ such a wealth of resources to improve their clinical care of patients. However, these clinical reports are of different quality, which needs to be selected.

It is important to check the study design, the statistical result and the conclusion. The P-value is the first value to be checked, to see whether the new proposition is confidently established against the null hypothesis. Effect sizes, such as odds ratios and hazard ratios, will then be checked to see how strong the observed effect is. It is important to discover whether the new proposition is validated in adequate studies. Finally, the performance indexes can be used to gauge whether the reported effect does exist, but also demonstrate the clinical usefulness of classifying patients.

6.5 Case studies

6.5.1 Exploring genomic variants associated to hepatitis C treatment

Chronic hepatitis C is a global pandemic disease affecting 170 million persons worldwide. The hepatitis C virus HCV can be classified as type 1 virus and non-type 1 virus. The former is more difficult to treat. Currently, the standard therapy is a 48-week course of pegylated recombinant interferon subcutaneous injection, combined with the oral antiviral agent Ribavirin. The treatment response is evaluated based on DNA viral load after the treatment course and a 24-week follow-up period. However, treatment responses manifest a great diversity among patients and across populations. Considering the synergistic antiviral effect by the host immune activity and the drug mechanism, the human DNA genotypes may be an important indication to the variability of treatment responses. It would therefore be justifiable to investigate the genome-wide associations between host (human) genotypes and the clinical outcome of standard therapy, a typical Pharmacogenomics study.

Two recent reports, published online in August and September 2009, respectively, represent two success stories of GWAS in this direction. The first report was an achievement of collaboration between Duke University, Schering–Plough Research Institute and Johns Hopkins University (Ge et al., 2009). A cohort of 1671 subjects was recruited from three ethnic groups: the European American, the African American and Hispanics. The subjects were classified into two groups, the sustained virological response (SVR) group and the Non-SVR group, based on the serum viral load detected at the end of the follow-up time. Genome-wide genotypes were obtained using the Illumina Human610-quad BeadChip.

Published by Woodhead Publishing Limited, 2013

A series of quality filters, primarily based on SNP call rates, were employed to safeguard the data quality.

The SNP rs12979860, residing in the 3kb upstream region of the IL28B gene (also known as Interferon lambda), is strongly associated to the treatment response. The significance level of association (P-value) is 1.37×10^{-28}, which is based on the trend test calculated by the logistic regression method. Comparing subjects with the risk genotypes CT or TT vs. those of the protective genotype CC in the European-American cohort, the (genotype based) odds ratio of being Non-SVR is 7.3.

Despite the varied allele frequency among different ethnic groups, people with the risk genotypes CT and TT consistently show lower rates of SVR, compared with people with the protective genotype CC. The consistency of association across populations demonstrates that the detected SNP reflects an underlying biological mechanism occurring in all these populations. This suggests the legitimate use of cohorts with mixed population for the detection of biological effects.

However, the authors also observed counter-intuitive findings that the protective CC allele is also associated positively to a higher baseline viral load. Intuitively, higher baseline viral loads represent more serious disease. The contradictory results show that the complex underlying mechanism defies simple explanations.

The second study was reported by Japanese scientists Tanaka et al. (2009). They recruited 142 Japanese subjects infected by Type I HCV for a genome-wide screen. Among them, 64 subjects had null virological response and 78 subjects had virological response. Here null virological response was defined as with less than a 2-log-unit decline in the serum HCV RNA level from the baseline value within the first 12 weeks and with detectable viremia in the 24th week after treatment. Affymetrix SNP 6.0 arrays were used for the genome-wide association to the therapeutic response. They further recruited a cohort of 172 subjects for the validation on a focused set of SNPs. The original GWAS data were filtered by quality filters mainly based on call rates, as well as the Hardy–Weinberg equilibrium tests in the virological response group.

Several SNPs in the IL28 gene were once again detected. The leading SNP rs8099917 (T/G) were detected with the minor allele G being the risk allele of null response. Comparing GG plus GT vs. TT in the genome-wide screening cohort, the chi-square test showed a P-value of 3.11×10^{-15}. The (genotype-based) odds ratio is 30.0.

The repetitive detections of SNPs in the same IL28 region, with different study cohorts across several populations, offer a mutual validation. It is now substantiated that SNPs in the IL28B gene can serve as a predictive biomarker for the treatment efficacy of peg-interferon and ribavirin combination on HCV patients. These studies unveil the role of host immune genes on the antiviral effect, and also demonstrate the credibility of the GWAS approach for pharmacogenomics studies. Most importantly, this test has now been commercialized for clinical use.

Despite the invention of several nucleoside analog anti-HBV drugs, Interferon still remains an important medication for treating HBV infected patients. Interferon is an endogenous innate defense agent against virus infection. The administration of Interferon can theoretically boost the immune system to fight against the viruses. Compared with various nucleoside-analog drugs, Interferon is less likely to invoke drug resistance capability of the viruses.

However, the responsiveness of Interferon for treating HBV patients remains uncertain, because patients usually have to be treated for several months before knowing whether the drug is effective. This motivates us to study the genetic associations to drug responsiveness.

6.5.2 Genetic tests for the adverse effect of Warfarin

Warfarin is the most widely used anticoagulant, which is used to decrease blood clot formation in human blood vessels. It has a variety of medical uses such as in the prevention of clots in patients who have implanted artificial heart valves, the treatment of deep vein thrombosis, pulmonary embolism, and antiphospholipid syndrome.

For years it has been difficult for doctors to give proper doses of Warfarin. If too much of the drug is given, the risk of hemorrhagic stroke and heavy internal bleeding is increased. If too little is given, the expected anticoagulation effect cannot be achieved. Yet, due to the personal difference in the metabolism of this drug and the response to the drug effect, the appropriate dose of Warfarin can vary by a factor of 10 among patients. Until now doctors adjust the dose of Warfarin by their experience and by patients' responses.

It has been gradually understood that many genomic variants in a gene responsible for the metabolism of Warfarin, namely CYP2C9 (cytochrome P450, family 2, subfamily C, polypeptide 9), and in a gene related to the drug response, such as VKORC1 (vitamin K epoxide reductase complex, subunit 1), are associated to the disease (Yuan et al. 2005). Repeated studies have demonstrated the associations with small P-values (i.e. the association is not likely to be false positive). The challenge for the clinical

Published by Woodhead Publishing Limited, 2013

use of this piece of knowledge is defined as a prediction equation to identify the optimum dose for each subject.

A recent study was conducted to meet such a challenge (Klein et al., International Warfarin Pharmacogenetics Consortium, 2009). This study enrolled 4043 subjects for building the prediction equation (80% of all subjects), and used 1009 subjects for validation of this equation (20%). The best dose for each subject was known and used to train the parameters of the equations. The R^2 was used to test the accuracy of prediction. A Warfarin dose algorithm based on several clinical and genetic variables such as height, weight, age and genotypes was developed. This algorithm is more precise than the clinical algorithm and fix-dose approach, especially for the population that required 21 mg or less of Warfarin per week or 49 mg or more per week for therapeutic anticoagulation. Besides, this genetic test result can tell us why the average Warfarin dose for Asians is lower than the dose for Europeans.

6.5.3 Tests for the need of preventive treatment after surgery

The Oncotype test aims to estimate the probability of post-operative distance recurrence (i.e. metastasis) in women with early stage breast cancer. It was developed against the background that preventive chemotherapy (i.e. Tamoxifen) was prevailingly prescribed to breast cancer patients after surgical treatment, for the sake of avoiding distant recurrence. The probability of post-operative distant recurrence with Tamoxifen treatment is about 15% in 10 years. It is thus suspected that not every patient will have recurrence and require preventive chemotherapy. A reliable predictive test is thus in great demand to indicate whether the preventive medication is required for a particular patient.

Unlike the two aforementioned tests, which are based on genomic (DNA) information, the Oncotype test relies on RNA information. A reverse transcriptase polymerase chain reaction (RT-PCR) method was developed to assess RNA expressions on surgical tumor tissues. The expression measurements of a panel of 21 genes (16 cancer-related genes and 5 reference genes) were combined using a regression model, and the estimated probability was given by a recurrence score between 0 and 100. This model was first developed on 447 subjects, then prospectively validated on 668 subjects (Paik et al., 2004). Due to the relative nature of RNA measurements, the reference genes were selected for normalization purpose. Expression levels of cancer-related genes were first divided by those of reference genes to provide a comparable measurement across subjects. The normalized measurements were basically ratios, which then were taken logarithmically before entering into the equation.

The Oncotype test is commercialized by the company of genomic health (Nasdaq: GHDX). The list price for a test is in the range of $3200 to $3500. The Oncotype test has been on the market since 2004. The revenue of Oncotype DX was $5.2 million in 2005 and $29.2 million in 2006. Major patents were granted in 2006 and 2007. A new test for colon cancer has been available on the market since 2010.

6.6 Take home messages

- Diagnostic biomarkers reflect the "status quo", while predictive biomarkers reflect the probability of future events.
- Different omics platforms are required for different goals. Case vs. control comparison of gene expression usually cannot find predictive biomarkers.
- Regression based methods are an important class of multivariate analyses, which jointly analyze multiple factors.
- Performance indexes include sensitivity, specificity, PPV and NPV, and the requirement depends on applications.
- Study design, sample size, P-values, effect sizes and validation studies are important aspects to evaluate the information in published literature for enhancing clinical practice.

6.7 References

Ge, D. et al. (2009) Genetic variation in IL28B predicts hepatitis C treatment-induced viral clearance. *Nature*, **461**: 399–401

Klein, T.E., Altman, R.B., Eriksson, N., Gage, B.F., Kimmel, S.E., et al. (2009) International Warfarin Pharmacogenetics Consortium. Estimation of the Warfarin dose with clinical and pharmacogenetic data. *N. Engl. J. Med.*, **360**(8): 753–64.

Paik, S., Shak, S., Tang, G., et al. (2004) A multigene assay to predict recurrence of Tamoxifen-treated, node-negative breast cancer. *N. Engl. J. Med.*, 351(27): 2817–26.

Published by Woodhead Publishing Limited, 2013

Patterson, S.D., et al. (2011) Prospective-retrospective biomarker analysis for regulatory consideration: White paper from the industry pharmacogenomics working group. *Pharmacogenomics*, **12**(7): 939–51.

Pearson, H. (2009) Human genetics: One gene, twenty years. *Nature*, 460(7252): 164–9.

Peduzzi, P., Concato, J., Kemper, E., Holford, T.R. and Feinstein, A.R. (1996). A simulation study of the number of events per variable in logistic regression analysis. *J. Clin. Epidemiol.*, **49**(**12**): 1373–9

Rommens, J.M., Iannuzzi, M.C., Kerem, B., et al. (1989) Identification of the cystic fibrosis gene: chromosome walking and jumping. *Science*, **245(4922)**: 1059–65.

Saad, E.D. and Katz, A. (2009) Progression-free survival and time to progression as primary end points in advanced breast cancer: often used, sometimes loosely defined. *Ann Oncol.* **20**(3): 460–4.

Tanaka, Y., Nishida, N., Sugiyama, M., et al. (2009) Genome-wide association of IL28B with response to pegylated interferon-α and ribavirin therapy for chronic hepatitis C. *Nature Genet.*, **41**(**10**): 1105–09.

Yuan H.Y., Chen, J.J., Lee, M.T., et al. (2005) A novel functional VKORC1 promoter polymorphism is associated with inter-individual and inter-ethnic differences in warfarin sensitivity. *Hum Mol Genet.* **14**(**13**): 1745–51.

7

Conclusions

DOI: 10.1533/9781908818232.147

Abstract: Conducting contemporary biomedical science is an art. Akin to food preparation, the bioinformatician needs to minimize the alteration of data (the food ingredients), but maximize its potential new knowledge and the visual presentations to biology (taste, flavor and aroma). Conceptual frameworks and adaptive computational models are two resources for elevating biomedical science to new heights.

Key words: conceptual framework, adaptive model.

7.1 Change and move forward

In current times, we are surrounded by fast advancing technology, such as smartphones and electronic commerce, which is drastically changing our lifestyle. In contrast, the general public perceives the improvements in medicine as relatively slow. Many unmet medical needs still remain, particularly in the treatment of serious diseases. Despite daily progress in basic biomedical science, lack of ground-breaking clinical applications prevents the lay person from perceiving any progress. The chasm between public expectations and scientific progress has developed into:

1. scientists being aware of their social responsibilities and devoting more effort outside their lab to present their findings to the general public;
2. the employment of a more global, holistic view and utilization of more molecular details for addressing higher level medical questions pertinent to our daily life;

3. the translation of biomedical knowledge from bench to bedside for creating clinically useful applications.

7.2 Presentation, presentation, presentation

The analysis of data of complex biomedical systems is akin to cooking, where the data are raw ingredients. The purpose of cooking is to release and balance the original taste and flavor of the ingredients, by food preparation and seasoning. It is the ingredients (i.e. the data) that carry the important biological clues. The responsibility of the bioinformatician is to present the clues truthfully, while maximizing presentation. We have discussed many useful presentations such as the Manhattan plot, the volcano plot, the Kaplan–Meier plot, the principal component plot and multi-scale genome browsers. These skills can be used flexibly to present the data to convey the message to the audience.

7.3 Domain knowledge plus adaptivity

With the availability of high-throughput technologies and fast accumulating data, it is imperative to increase the throughput of knowledge generation and clinical applications, based on a high concentration of insights. In different parts of this book, we show that adaptive models are critical for the exploration of a deluge of data scattered in a wide and non-linear search space. Examples of adaptive models include the genetic algorithm used for genomic variant analysis; k-means clustering for forming gene modules from RNA level data; principal component analysis for the analysis of sample clustering; and swarm intelligence for systems biomedical science. By letting the data speak for themselves, we can circumvent the limitations resulting from our linear thinking, distil the essence of matter, and achieve meaningful results in a shorter time.

The fairytale of *Snow White and the Seven Dwarfs*, told by the Grimm Brothers, is a well-known bedtime story. The queen often asked the magical mirror: "Mirror, mirror on the wall, who in the land is the fairest of all?" When we are confronted with the messy, tremendous volumes of biomedical data, we share the same feeling and are urged to ask the

Published by Woodhead Publishing Limited, 2013

computer agents: "Computer, computer, when can the hidden valuable knowledge emerge from the data?" We hope that the computer is wise enough to give us a crisp and insightful answer after intensive computation of the messy data. But most of the time, a computer agent is more like a hard working but relatively inexperienced student who needs to be coached. A computer agent is willing to work even in the middle of the night, yet has difficulty in telling apart trivial observations from real insights. At the end of the day, we still need to provide guidance and directions to computer agents, so as to extract new knowledge from the data.

Index

Published by Woodhead Publishing Limited, 2013

CPSIA information can be obtained at www.ICGtesting.com
Printed in the USA
LVOW07*1427140813

347902LV00009B/142/P